MW01324984

Important People *in* Biotechnology

Titles in the series:

Words *and* Terms
Important People *in* Biotechnology
The History *of* Biotechnology
Debatable Issues

A STUDENT'S GUIDE TO

biotechnology

VOLUME 2

Important People *in* Biotechnology

GREENWOOD PRESS
Westport, Connecticut · London

Library of Congress Cataloging-in-Publication Data

Creative Media Applications
A student's guide to biotechnology/Creative Media Applications.
v. cm.
Includes bibliographical references and index.
Contents: v.1. Words and terms — v.2. Important people in biotechnology — v.3. The history of biotechnology — v.4. Debatable issues.
ISBN 0-313-32256-2 (set) — ISBN 0-313-32257-0 (v. 1) — ISBN 0-313-32258-9 (v. 2) — ISBN 0-313-32259-7 (v. 3) — ISBN 0-313-32260-0 (v. 4)
1. Biotechnology — Juvenile literature. [1. Biotechnology.] I. Creative Media Applications.

TP248.218.S78 2002
660.6–dc21 2002072693

British Library Cataloguing in Publication Data is available.

Library of Congress Catalog Card Number: 2002072693
ISBN: 0-313-32256-2 (set)
0-313-32257-0 (Vol. 1)
0-313-32258-9 (Vol. 2)
0-313-32259-7 (Vol. 3)
0-313-32260-0 (Vol. 4)

First published in 2002

Greenwood Press, 88 Post Road West, Westport, CT 06881
An imprint of Greenwood Publishing Group, Inc.
www.greenwood.com

Printed in the United States of America

The paper used in this book complies with the Permanent Paper Standard issued by the National Information Standards Organization (Z39.48–1984).

10 9 8 7 6 5 4 3 2 1

A Creative Media Applications, Inc., Production
WRITER: Andy McPhee
CONTRIBUTING EDITOR: Joanne Poeggel
DESIGN AND PRODUCTION: Fabia Wargin Design, Inc.
EDITOR: Matt Levine
COPYEDITOR: Laurie Lieb
ASSOCIATED PRESS PHOTO RESEARCHER: Yvette Reyes
CONSULTANTS: Thomas Shannon, Worcester Polytechnic Institute
Cathy Shannon, Reading Recovery Teacher, Worcester MA

PHOTO CREDITS:
Cover: Photodisc, Inc. & AP/Wide World Photographs
AP/Wide World Photographs *pages:* vii, 8, 12, 20, 25, 26, 35, 38, 43, 63, 74, 90, 93, 100, 107, 112, 119, 120, 123, 124, 125
Courtesy of the Cold Spring Harbor Laboratory Archives *pages:* 5, 47, 65, 77, 114
©CORBIS *page:* 68
©Bettman/CORBIS *pages:* 2, 14, 33, 45, 51, 52, 54, 56, 59, 71, 81, 85, 87, 95, 96, 103, 105, 116, 123
©Hulton-Deutsch Collection/CORBIS *pages:* 17, 122
©Ted Streshinsky/CORBIS *page:* 23
©Digital Art/CORBIS *pages:* 29, 124
©Roger Ressmeyer/CORBIS *page:* 109

Table of Contents

Introduction

Biotechnology (*bye*-oh-tek-**nahl**-oh-jee)—the use of living things to make an impact on other living things—includes many areas of science. When a new medicine is created, or when a clone (klohn) of a sheep is made from the cells (selz) of another sheep, that is biotechnology. Using genetics (juh-**net**-iks) to combine the best *traits* (traytz)—specific qualities or features—from two different types of plants to make new crops that taste better, resist disease, or stay fresh longer involves biotechnology.

Biotechnology has been one of the most exciting, fastest-growing sciences of the last century. It promises to make even more astounding advances in this century. The people who have been involved in this exciting enterprise are as complex and fascinating as the science itself. People involved in biotechnology come from the fields of biology, medicine, genetics, chemistry, industry, and agriculture.

For those interested in the human side of biotechnology, here is a collection of some of the most notable people involved in this field through the centuries. People included in the collection made important contributions to advancing scientific knowledge. Their work required them to use living organisms (**or**-gan-*iz*-umz) to change other organisms or to create new organisms. (An *organism* is any living thing.) They come from different times and places and are different in age, gender, personality, and cultural background. They were curious to understand the world around them. They were well educated—in some cases, self-educated. All were persistent and continued to question, explore, and experiment despite setbacks. Their successes put humankind on a path to a greater understanding of the world.

These biographies don't constitute a complete list. Many other people might be included here, all of them important in their own ways. By reading the biographies in this collection,

Sherrill Neff (left), president and CFO of Neose Technologies, and Stephen Roth, chairman and CEO, pose at the company lab in Horsham, Pennsylvania, in August 1999. Neose Technologies is a biotechnology company that uses computer research and genetic engineering to search for new and better medicines.

you will be better able to understand biotechnology and how the advances that now seem so common were first achieved.

Pronunciation

You will notice that in the pronunciation guides, the words are not broken into syllables. Rather, they are sectioned off by the way that they sound so you can figure out how to say them. The piece of each word in **bold type** is where you put the greatest emphasis. That means it is the part you pronounce slightly louder.

The piece of a word in *italics* is where the second emphasis is. When you say the word, the part in italics is pronounced softer than the part in bold but louder than the other syllables of the word.

Note: All metric conversions in this book are approximate.

Oswald Avery (1877–1955)

Oswald Theodore Avery was a research physician and *bacteriologist,* a scientist who studies bacteria (bak-**teer**-ee-uh). *Bacteria* are one-celled organisms that can be seen only with a microscope. *Cells* are the smallest units of a living thing that can grow, reproduce, and die. Avery discovered that *deoxyribonucleic* (dee-*ox*-ee-rye-boh-noo-**klay**-ik) *acid,* or DNA, makes up a cell's genes (jeenz). *Genes* determine the nature of every cell in the body and how it will function. With this discovery, Avery laid the foundation for modern genetic research.

Avery was born on October 21, 1877, in Halifax, Nova Scotia, Canada. His father, a Baptist minister, was invited to become pastor of a church in New York City. In 1887, his family moved there.

Avery graduated from Colgate University in 1900. He then attended the Columbia University College of Physicians and Surgeons and became a doctor. After only three years, however, Avery became frustrated by his work. He believed that science was limited in its ability to help patients. He thought that he might be of more help to patients by doing medical research.

In 1907, Avery went to work for the Hoagland Laboratory in Brooklyn, New York. The laboratory, the first of its kind in the country, specialized in studying bacteria. At the lab, Avery began studying the chemicals in bacteria, hoping to understand exactly how they caused disease.

Avery was invited to join the Hospital of the Rockefeller Institute for Medical Research in New York City in 1913. This hospital allowed researchers to work in its labs and use its patients to study diseases firsthand. One of the diseases that they studied was *pneumonia* (noo-**mohn**-yuh), a lung infection. Avery joined the pneumonia research team, and for the next thirty-five years, his work focused on a single type of pneumonia-causing bacteria—*pneumococci* (*noo*-moh-**kahk**-*sye*).

Oswald Avery's years of work with pneumococci bacteria were a key step in the founding of molecular biology.

Avery's work on pneumococci continued throughout World War I (1914–1918). During the war, Avery was a captain in the U.S. Army. His wartime duties included teaching army medical officers how to treat pneumonia.

After the war, Avery went back to his research at the Rockefeller Institute. By 1940, he began to wonder why one type, or *strain,* of pneumococci didn't cause serious infection and possible death, and the other strain did. He found that the non-fatal strain contained a *mutation* (myoo-**tay**-shun), a change in a single gene that can be passed on to offspring. (*Offspring* are young organisms produced by parents.) When a colleague performed an experiment that mixed the two kinds of pneumonia bacteria—the mutant strain and the deadly strain—the mutant strain also became deadly.

Avery was curious to know why that had happened. He and his research team studied tubs of the bacteria. Avery eventually concluded that the deadly bacteria were able to pass their traits along to the mutant bacteria through something now known as DNA.

Avery outlined the results of his research in a paper published in 1944. Other scientists doubted Avery's findings. They believed that an unknown protein (**proh**-teen) might have caused the change in the bacterial strain. (*Proteins* are substances essential for all living cells.) Even so, Avery's paper led to more studies of DNA. These studies eventually showed that DNA held genetic information and was present in all animal cells.

Avery continued his research at the Rockefeller Institute until 1948, when he retired to Nashville, Tennessee. He was found to have liver cancer in 1954 and died about a year later, on February 20, 1955, at the age of seventy-seven.

Paul Berg (1926–)

Paul Berg is considered one of the fathers of genetic engineering (*en*-jin-**eer**-ing). *Genetic engineering* is the process of changing genes in an organism to alter the characteristics (*kaehr*-ik-tuh-**riss**-tiks) of offspring. *Characteristics* are features inherited (in-**herr**-it-ed) by offspring; to *inherit* is to receive from one's parents.

In 1972, Berg combined DNA from a cancer-causing virus (**vye**-rus) found in monkeys with that of another virus to create the first recombinant (ree-**kahm**-bih-nunt) DNA molecule (**mahl**-uh-*kyool*). A *virus* is a tiny structure that can live in a cell and cause disease. *Recombinant DNA* is DNA formed by splitting a DNA molecule from one organism and combining it with DNA from another organism. A *molecule* is the smallest bit into which a substance can be divided without chemical change. Berg was also the first person to make a human hormone (**hor**-mohn) by combining a virus with genes from bacteria. *Hormones* are naturally occurring chemicals that are released by one part of the body but have an effect somewhere else.

Berg was born in Brooklyn, New York, and attended Western Reserve University, now Case Western Reserve, where he received his Ph.D. in biochemistry (*bye*-oh-**kem**-ih-stree) in 1952. *Biochemistry* is the study of the chemistry, or structural makeup, of living things. He became a professor of biochemistry at Stanford University School of Medicine in Palo Alto, California, in 1959, when he was thirty-three. By the time that he turned forty, he was elected to the National Academy of Sciences, a group of scientists and researchers dedicated to advancing science and technology.

After his development of recombinant DNA techniques in 1972, Berg became concerned about the danger of accidentally creating a new disease. He stopped his experiments and sent a letter to the scientific community about his concerns. The letter is now referred to as the "Berg letter." In it, he proposed that scientists stop recombinant DNA research for one year. He wanted to make sure that new diseases wouldn't be released into the world.

Berg was a key organizer of a world-famous conference about recombinant DNA technology. This conference, the Asilomar Conference, was held in California in February 1975. One hundred leading scientists met there to discuss the risks of genetic experiments. The issues discussed at the conference resulted in a set of guidelines for DNA research, published by the National Institutes of Health (NIH) a year later. (The NIH is a collection of government agencies that conduct research and set national policies about all aspects of health care.) The guidelines show how scientists should regulate their own research.

Paul Berg was a great pioneer in genetic engineering.

Berg eventually continued his recombinant DNA research. In 1980, he was awarded the Nobel Prize in Chemistry for his work. (The Nobel prize is an international award given every year for important research in physics, chemistry, medicine, literature, and peace.) Berg's laboratory continued to work with recombinant DNA techniques throughout the 1980s.

Berg became director of the New Beckman Center for Molecular and Genetic Medicine in Stanford, California, in 1985. He held that position until 1991, when he accepted a position as head of the Scientific Advisory Committee of the Human Genome (**jeen**-ohm) Project, a position that he still holds. *The Human Genome Project* is composed of an international team of scientists whose goal is to identify the full set of genetic instructions contained inside each human cell. The project consists of two phases. The first phase, to create a complete map of all genes in the human body—the human *genome*—was completed in 2000. This enormous undertaking included the identification of some 30,000 genes. This number may change as more precise research is done.

All genes appear in the same order in every human chromosome (**kroh**-muh-*sohm*). A *chromosome* is a long, threadlike strand of DNA in a cell's *nucleus* (**noo**-klee-us), the activity center of the cell. About 99 percent of human genes are the same from person to person. The 1 percent that are different make each person an individual. When the map of the genome was completed, President Bill Clinton (1946–) said, "This is the most important, most wondrous map ever produced by humankind."

The second phase of the project involves determining how the chemicals in DNA are organized into genes and how those genes affect the body. This phase is expected to take about fifteen years and will require dedication on the part of all scientists involved. Berg has shown his dedication many times during his career. He hopes more young people become scientists. "Great scientists," says Berg, "like great athletes, have to have a passion, to be driven to discover and solve problems."

Norman Borlaug (1914–)

Unlike most well-known scientists, Norman Borlaug won most of his fame outside the laboratory. Once he perfected a new strain of wheat, Borlaug traveled to farms in Asia and Africa, introducing it to a hungry world. Borlaug has been called the father of the Green Revolution, an effort to increase crop yields and introduce new crops in poor nations. For his efforts, he won the 1970 Nobel Peace Prize. In a 1997 profile of Borlaug for *The Atlantic Monthly,* journalist Gregg Easterbrook wrote, "Though barely known in the country of his birth, elsewhere in the world Norman Borlaug is widely considered to be among the leading Americans of our age."

Borlaug was born in Cresco, Iowa, in 1914. As a boy, he worked on his family's farm. He attended the University of Minnesota, earning a degree in forestry in 1937. During his college years, the United States was experiencing the Great Depression, a time when a bad economy left millions of Americans out of work, homeless, and hungry. The country also went through a severe drought that turned many farmlands on the Great Plains to dust. Borlaug noted, however, that some farms were not as harshly affected. These tended to be farms whose owners practiced *high-yield agriculture*—growing the most crops possible in a small area of land.

Borlaug eventually earned advanced degrees in plant pathology, or disease, and then took a job with DuPont, a chemical company. He helped research chemicals to kill microorganisms that can damage plants. In 1944, he went to Mexico to work as a geneticist for the Cooperative Wheat Research and Production Program. The program focused on studying soils, finding new strains of wheat, and fighting diseases that affected the crops.

As a geneticist (juh-**net**-uh-sist), Borlaug drew on others' earlier work in breeding new crops. A *geneticist* is a scientist who works in the field of genetics. The roots of this science—the study of how organisms pass on genes to offspring—go back to the

Iowa governor Tom Vilasck (right) looks on as Dr. Evangelina Villegas receives a Millennium World Food Prize in October 2000. Presenting the award is Norman Borlaug, whose efforts at creating crops to solve world hunger earned him the 1970 Nobel Peace Prize.

Austrian monk Gregor Mendel (1823–1884). Mendel did his work with pea plants, but later scientists used corn and wheat.

U.S. scientist George Harrison Shull (1874–1954) worked with corn. Shull studied traits in corn that were easy to record, such as how many kernels a corncob produced. Corn and other plants reproduce when pollen enters their flowers. *Pollen* is made up of tiny grains from seed-producing plants. Usually yellow in color, pollen grains can also be red, white, brown, or purple. Pollen from one plant that lands on another plant may *pollinate* the second plant, meaning that male pollen joins with a female cell in the plant. Pollination results in the production of seeds.

Shull took a corn plant's own pollen and put it into one of its flowers to create offspring. He repeated this process again with the offspring to create the next generation, a process called *inbreeding*. Each generation would have, on average, the same number of kernels.

Inbreeding, however, led to a problem: The process weakened future generations of corn, limiting their growth. Shull's solution was to breed one inbred type of corn with another—a process called *crossbreeding*. In crossbreeding, farmers plant crops near other, related crops to encourage *cross-pollination* (kross pahl-in-**ay**-shun), which occurs when pollen grains released by the flower of a plant are carried to other plants by wind, insects, birds, or other animals. The offspring of Shull's two inbred corns had the traits of the parent corn but grew better than the inbred lines. The offspring of the two inbred lines was called a *hybrid* (**hye**-brid).

By Borlaug's time, scientists had perfected crossbreeding so that the traits of four different inbred lines of a crop could be combined in a hybrid. Borlaug's focus was on wheat. He wanted to create a strain that would produce more grain. His work led to a wheat plant with a short stalk, called *dwarf spring wheat*. The plant was bred to use less energy growing its stalk and more energy forming grain. The new hybrid was also able to resist diseases that killed other types of wheat. The improved wheat led to greater harvests per each acre planted, compared to older wheat strains. In other efforts, Borlaug developed new crops that could grow well regardless of the amount of sunlight that they received each day. He also helped perfect *triticale,* a hybrid of wheat and rye often used to feed farm animals.

After proving the success of his wheat hybrid in Mexico, Borlaug wanted to use it in other countries facing rapid population growth. By the early 1960s, India and Pakistan seemed close to massive starvation, because their farmers could not produce enough food to feed the growing number of citizens. In 1965, Borlaug convinced the two countries to let him grow his dwarf spring wheat there. Raising the crop required fertilizers (**fur**-tih-*lyze*-urz) that had never been used in these countries before. (*Fertilizers* are nutrients added to the soil to produce bigger and better plants.) This high-yield farming led to a dramatic rise in the production of wheat and other crops. Within several years, Pakistan was growing enough wheat to feed all its people. By the 1980s, India had enough wheat to sell to other countries. By using Borlaug's hybrid crops and his farming methods, the countries did not have to cut down forests to create new farmland.

Despite his success in Asia, Borlaug had his critics. By the 1980s, some people were attacking high-yield agriculture. The process required too much fertilizer, they said, which was often made from chemicals that posed health risks to people. Other critics said that farmers around the world should grow foods native to their own countries, not Borlaug's hybrid wheat.

Borlaug had hoped to bring the Green Revolution to Africa. However, charity groups that had supported his earlier efforts stopped providing funds due to pressure from critics. He retired briefly, but then renewed his efforts when he found a new source of money. A Japanese millionaire offered funds so that Borlaug could go to Africa. Borlaug's methods led to a rapid increase in the production of such crops as corn, wheat, and sorghum, another grain.

Now in his eighties, Borlaug remains active in the efforts to reduce international hunger, and he teaches at Texas A&M University. In addition to his Nobel prize, he has received many other awards for his work. Borlaug also awards his own honor, the World Food Prize, to people who promote increased food production and the end of hunger.

Herbert Boyer (1936–)

As a boy, Herbert Boyer dreamed of playing professional football. He played for his high school team in Derry, Pennsylvania. Boyer's coach was also one of his science teachers, and he convinced Boyer that science could be just as rewarding as sports. Football's loss was Boyer's—and the world's—gain, as he became one of the pioneers of modern biotechnology.

After graduating from high school, Boyer studied medicine at nearby St. Vincent's College. He found that he was not suited to be a doctor, so he studied biology and chemistry. He graduated from St. Vincent's in 1958 and went on to graduate school at the University of Pittsburgh, Pennsylvania, in 1963. Next, he spent three years at Yale University in New Haven, Connecticut, where he studied biochemistry and other sciences. Boyer became an assistant professor at the University of California at San Francisco in 1966.

In 1969, Boyer was studying a common bacterium, *Escherichia coli* (*eh*-shur-**ik**-ee-uh **koh**-lye), or *E. coli,* which is normally found in the intestines. During that research, Boyer made a key discovery in genetic engineering. He had been working with chemicals called *restriction enzymes* (**en**-zymez), used to cut DNA for study. Boyer saw that some restriction enzymes were able to cut DNA in a way that left what he called "sticky ends" on the DNA strands. These sticky ends could be used to connect pieces of DNA to each other.

At around the same time, Stanley N. Cohen (1935–) had discovered a way to remove plasmids from cells and insert them into other cells. *Plasmids* are circular strands of DNA typically found in bacteria, but also found in some cells. Plasmids carry genetic information from one bacterium or cell to another. Cohen, like Boyer, worked with *E. coli* in his research.

In 1972, Cohen and Boyer met at a conference in Hawaii. The two scientists discussed their shared interests at a now-famous meeting at a deli near Waikiki Beach. They began working together the next year. They used enzymes and plasmids to combine sections of DNA and insert them into bacterial cells. The offspring of the bacteria were then able to make certain proteins. Boyer's enzyme work combined with Cohen's plasmid studies to form the basis of the biotechnology industry.

Jeremy Rifkin, writing for the magazine *Business Week,* says that the discovery was almost as important as learning to control fire was for primitive humans: "[Cohen and Boyer] performed a feat in the world of living matter that some biotech analysts believe rivals the importance of harnessing fire.... [It] is a kind of biological sewing machine that can be used to stitch together the genetic fabric of unrelated organisms."

In 1975, Boyer met Robert Swanson (1947–1999), a businessman, who talked to Boyer about the possibility of using cells as factories to make medicines. In 1976, the two men created Genentech, a company dedicated to the new field of developing *biopharmaceuticals* (bye-oh-*far*-muh-**soo**-tih-culz), drugs produced by genetic engineering.

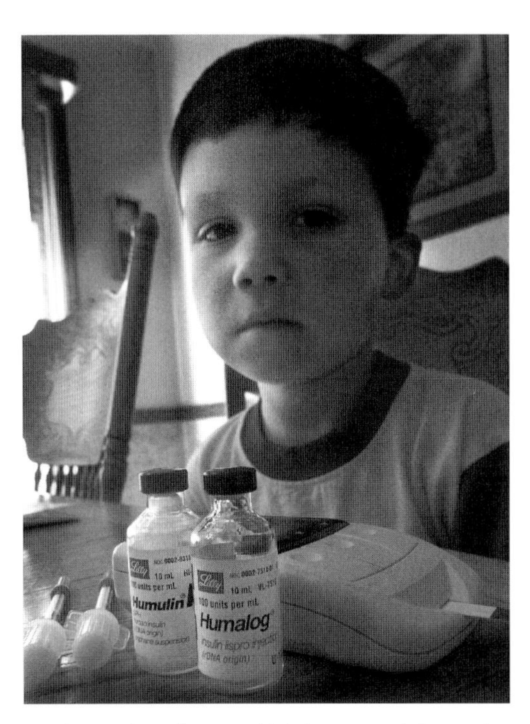

Andy Crowley, a five-year-old diabetic, sits with his insulin and syringes. Diabetics need regular doses of insulin to live healthy lives.

Scientists at Genentech first developed genetically engineered insulin. *Insulin* is a naturally produced hormone that allows body cells to use sugar for energy. In a person with the disease *diabetes,* either the body doesn't make enough insulin, or the insulin that it makes doesn't work as well as it should. Many diabetics need to inject themselves with insulin to survive.

Prior to Genentech's development, insulin was produced primarily from pigs. The pig form of insulin caused allergic reactions in some people. An *allergic reaction* is a response by the body to something that it considers foreign. Genentech set about to develop a human form of insulin that would have less risk of allergic reaction and fewer side effects. In 1978, Boyer produced an artificial (*ar*-tih-**fish**-ul), or human-made, version of the gene that makes human insulin. The gene was inserted into bacteria, causing the bacteria to produce human insulin.

Genentech scientists later created many other biopharmaceuticals, including *human growth hormone* (used to treat patients with certain growth disorders) and *tissue plasminogen* (**tish**-oo plaz-**min**-uh-jen) *activator,* or tPA (used to stop heart attacks from occurring). Since Boyer and Cohen's original work, more than 100 drugs have been manufactured through genetic engineering. An estimated 200 million people have had their lives extended or saved thanks to these drugs. At the Genentech Web site, Boyer offers this observation: "Biotech's first quarter century has been truly remarkable. As a science and an industry, biotechnology has delivered on its promise of improving patients' lives." Boyer remains active on a leadership committee at Genentech.

Luther Burbank (1849–1926)

Luther Burbank was at the forefront of what is now called *agricultural biotechnology*. Burbank was born in Lancaster, Massachusetts, in 1849. As a boy, he lived on a farm. When he was twenty-one, Burbank bought land near Lunenburg, Massachusetts, and began raising plants.

For centuries, farmers had been modifying crops through crossbreeding. In crossbreeding, farmers plant flavorful, hardy crops near other, related crops to encourage cross-pollination. Farmers hope that when they plant a crop with desirable traits near a crop with less desirable traits, the more desirable plants will pollinate the less desirable plants. The resulting seeds might then grow into plants that have the desirable traits.

In 1873, at the age of twenty-four, Burbank used crossbreeding to develop what came to be known as the Burbank potato. This large, hardy potato was considered much tastier

Luther Burbank used crossbreeding to create many varieties of fruits and vegetables.

than the smaller potatoes usually grown at that time. It became very popular and was eventually taken to Ireland. A deadly fungus had stricken potato crops in Ireland late in the summer of 1845. Potatoes were an important food source at the time. The following year, the fungus destroyed nearly every potato crop in the nation. As a result, a famine struck Ireland. A million people died from starvation. Burbank's potato helped Ireland recover from the famous potato blight. The Burbank potato was later developed into the Russet Burbank, which remains a key crop of Idaho potato farmers.

In 1875, Burbank moved to Santa Rosa, California, where he lived and worked for the next fifty years. He established a nursery garden and a greenhouse for his experiments. There, he improved many varieties of fruits and vegetables, including plums, prunes, and berries. For instance, he crossed a plum with an apricot to create a new fruit that he called the *plumcot.* He also created the *freestone peach,* a variety in which the peach fruit separates easily from the pit. Burbank also created several new strains of roses and a new cactus without thorns.

Burbank was an expert in cross-pollination techniques to produce new strains of plants. In fact, he developed more than 200 new varieties of plants. In 1893, Burbank published a book called *New Creations in Fruits and Flowers.* The book described hundreds of new plant *species* (**spee**-sheez), or groups of related organisms, that he had created through cross-pollination. Several religious groups criticized the book. They claimed that only God could create new plants and that Burbank should stop playing God and end his research. Despite this criticism, however, Burbank's book eventually made him famous.

Burbank died in 1926, at the age of seventy-seven. At the time, he had more than 3,000 experiments underway in which he was growing different varieties of plants from around the world.

Ernst Boris Chain (1906–1979)

Scientists rely on the work of other researchers to confirm their discoveries or add new information. The history of the drug *penicillin* (*pen*-uh-**sil**-in) reflects this idea of shared knowledge and pursuits. British scientist Alexander Fleming (1881–1955) discovered that the mold *Penicillium notatum* (*pen*-uh-**sil**-ee-um *noh-***taht**-um) kills bacteria that cause disease. However, it took the work of Ernst Boris Chain and his partner Howard Florey (1898–1968) to show how effective the drug made from this mold could be.

Chain, born in Berlin, Germany, was the son of chemist Dr. Michael Chain. As a boy, Ernst often visited his father's laboratory and the chemical factory he owned. Following in his father's footsteps, the younger Chain earned a degree in chemistry in 1930, though for a time, he considered becoming a concert pianist. After college, he did research on enzymes in both Germany and Great Britain. (*Enzymes* are protein molecules that start or speed up chemical reactions within organisms without changing themselves.) Considered a brilliant biochemist, Chain also had an interest in antibiotics.

In 1935, Chain began working at Oxford University in England, where he met Florey. Chain had already discovered a paper that Fleming had written about his attempt to isolate and produce penicillin. Florey had also heard about Fleming's efforts. However, Fleming could not produce enough of the drug to make effective studies, and he ended his work.

In 1939, Chain and Florey began a detailed study of penicillin. Some of their funds came from the Rockefeller Foundation, a U.S. fund for science and other research. Chain was in charge of growing the mold and creating the pure form of penicillin that could be used as a drug. He also discovered the chemical structure of penicillin. He and Florey then tested its antibiotic capabilities using what was called "the mouse protection test," a test that had already been used to develop chemical antibiotics known as *sulfa drugs*.

The work that Ernst Boris Chain did with penicillin earned him the Nobel Prize in Physiology or Medicine in 1945.

A group of mice were injected with the bacteria *Streptococcus* (*strep*-toh-**kahk**-us). Half of the mice were also given penicillin. The mice that received the antibiotic lived, while the others died. The two scientists were lucky that they had chosen mice for their research. For many animals, penicillin acts as a poison. If Chain and Florey had used such animals in their test, they would not have learned that penicillin was an effective antibiotic.

After their success with the mice, Chain and Florey decided to test penicillin on a human. Their test subject, a British police officer, had developed a serious infection after cutting himself while shaving. The scientists were still not able to produce large amounts of penicillin, so during the test, they removed the small amounts of penicillin that had entered the officer's urine in order to reuse the drug. The patient eventually died because of the limited amounts of penicillin on hand. Still, the antibiotic did improve his condition for a time, and the drug had not harmed him. Chain and Florey then conducted several more tests on humans that showed penicillin's power to heal. The new drug often worked when sulfa drugs did not.

By this time, Great Britain was involved in World War II (1939–1945). Government officials were interested in using penicillin to prevent deadly infections caused by bacteria entering wounds. The country's drug companies, however, did not have the resources to produce large amounts of penicillin, and British factories were the targets of German bombs.

Chain and Florey began working with U.S. companies, using methods that were used to brew beer to make penicillin from mold. The U.S. government provided money for this effort, since the United States had now entered the war and would need penicillin for its wounded soldiers. Researchers also found a new type of penicillin mold that could be used to produce more penicillin than *Penicillium notatum*.

During the war, penicillin saved many lives—it was used to treat such diseases as pneumonia, scarlet fever, strep throat, and blood poisoning. In 1945, Chain, Florey, and Fleming shared the Nobel Prize in Physiology or Medicine for their work with penicillin. Other scientists have since found new antibiotics, but penicillin still remains effective for treating some diseases.

After the war, Chain left England for Italy, where he studied antibiotics at the International Research Center for Chemical Microbiology. He later returned to England, where he ended his scientific career. In 1969, he was made a knight. Chain once wrote that he did not pursue the development of penicillin because he wanted to create a "miracle cure" for infections. Instead, he was merely interested in the science behind antibiotics. His success showed how the "pure" research done in laboratories can have important practical results for the world.

Stanley Cohen (1922–)

Biochemist Stanley Cohen was born in Brooklyn, New York, in 1922. His parents had come to America from Russia in the early 1900s. His father was a tailor, and his mother was a homemaker. They both had limited education, but they taught Cohen that a good education was important.

Cohen went to public schools in New York City. He later studied biology and chemistry at Brooklyn College. He was particularly interested in cell biology and the development of *embryos* (**em**-bree-oze), which are the unborn young of animals. Cohen has said, "I think my one insight into these problems was the recognition that much could be learned by the application of chemistry to biology."

Stanley Cohen is shown in his lab after winning the Nobel Prize in Physiology or Medicine in October 1986. He shared the prize with Rita Levi-Montalcini for their work with growth factors.

Cohen worked as a bacteriologist in a milk-processing plant until he could afford to go to graduate school. He graduated with a degree in zoology from Oberlin College in Ohio in 1945. He next received a doctorate degree from the University of Michigan in 1948. To earn that degree, he had to write a long and detailed research paper. His paper discussed earthworm metabolism. The term *metabolism* (muh-**tab**-uh-*liz*-um) refers to the chemical changes that occur in an organism. Of those days, he recalled, "I remember spending my nights collecting over 5,000 worms from the University campus green."

After graduation, Cohen worked at the University of Colorado, where he studied the metabolism of infants born earlier than expected. He next went to Washington University in St. Louis, Missouri. There, he and Rita Levi-Montalcini (1909–) isolated nerve growth factor, which Levi-Montalcini had found in certain mouse tumors. *Growth factors* are body chemicals that help control the growth of different kinds of cells and organs.

Cohen also discovered epidermal growth factor, which plays a key role in healing wounds. It is important for proper development of an embryo, as well. Cohen's discovery of epidermal growth factor led to more studies on how cells and tumors grow. It also led to the development of new treatments for many diseases.

In 1959, Cohen went to Vanderbilt University in Nashville, Tennessee, where he became a professor in the biochemistry department. He and Levi-Montalcini were recognized for their work with growth factors in 1986, when they jointly received the Nobel prize. Cohen retired in 1999, but he remains a professor emeritus of chemistry at Vanderbilt.

Stanley N. Cohen (1935–)

As a boy, Stanley N. Cohen thought about studying atomic physics, but a high school biology teacher introduced him to a new field: genetics. After that, Cohen devoted himself to the role that genes play in all creatures, from viruses to humans. Along the way, he helped spark a revolution in science.

Cohen was born in 1935 in Perth Amboy, New Jersey. He majored in biological studies at New Jersey's Rutgers University and studied medicine at the University of Pennsylvania, graduating in 1960. Cohen then turned to research, and by 1968, he was working at Stanford University, in Palo Alto, California. His focus was plasmids, circular strands of DNA that carry genetic information from one bacterium or cell to another.

Cohen wanted to learn how the genetic information in plasmids made bacteria resistant to some medicines. He worked with the common bacterium *E. coli,* which is normally found in the intestines. In 1971, Cohen and his assistants found a way to introduce a certain plasmid into *E. coli*. This plasmid carried DNA that made the bacterium resist the effects of a particular medicine. When the bacterium reproduced, its offspring also had the gene that resisted the drug.

Around the same time, Herbert Boyer (1936–), at the University of California at San Francisco, was working with *E. coli* as part of his study of chemicals called restriction enzymes. He used these enzymes to cut DNA in a way that left "sticky ends" on the DNA strands. These sticky ends provided a way to join pieces of DNA to each other.

In 1972, Cohen and Boyer met at a conference in Hawaii. In a now-famous meeting at a deli near Waikiki Beach, the two scientists discussed their shared interests. Cohen recalled during a 1999 lecture in San Francisco, "Over hot pastrami and corned beef sandwiches, I proposed a collaboration with Herb, and the two of us began to plan the experiments we would carry out."

Dr. Stanley N. Cohen's work with plasmids helped revolutionize the field of genetics.

Cohen and Boyer's historic work took place separately at their labs. One of Cohen's assistants took plasmids back and forth between Palo Alto and San Francisco. The scientists worked with *E. coli*. Using one of Boyer's enzymes, they cut a plasmid and added a new gene. Then they rejoined the cut plasmid. The offspring of the bacterium then produced certain proteins, proving that the gene modification was successful.

In further work, Cohen was able to take genes from one kind of bacterium and introduce them to different kind of bacterium. Then he took genes from more complex living organisms—first frogs and then mammals—and introduced them to bacteria. The genetically modified bacteria then served as living "factories" to make chemicals produced by the genes of the other creatures. Cohen and Boyer received a patent (**pat**-ent) for their process, which has earned several hundred million dollars for Stanford and the University of California. A *patent* is a legal document from the U.S. government giving an inventor the exclusive right to make, use, and sell an invention for a certain period of time.

In recent years, Cohen has focused on the role that genes play in cancer. He and his assistants have developed a process that can shut off the activity of certain genes in mammals. This enables them to see if "turning off" the genes has any effect on the growth of cancer cells in the organism. The process helps identify the genes that may fight cancer. This work could also help scientists learn which genes control resistance to the drugs used to treat many cancers.

Cohen will always be remembered as one of the most important figures in biotechnology. However, Cohen himself is quick to acknowledge all the researchers who paved the way before him. "Science isn't done in a vacuum," he said at his 1999 lecture. "The work I have talked about depended on the fruits of many years of effort in laboratories throughout the world.... One can say with some certainty that if Boyer and I had not collaborated...in early 1973, [the same advances] would have been achieved later by other labs."

Francis Collins (1950–)

Francis Collins is the current leader of the Human Genome Project. This international effort to map the human genome was partially completed on June 27, 2000. At that time, scientists finished creating a map of all currently known human genes, some 30,000 in all. The team is working now on the second phase of the project, to determine the precise sequence of the chemicals in every strand of DNA in a cell.

With Francis Collins (right) standing by him, President Bill Clinton announces the Human Genome Project's success in completing a rough map of all known human genes in June 2000.

Collins took over the role as leader in 1994 from James Watson (1928–). Before taking this position, Collins performed important research in identifying genes that cause disease.

Lab assistant Becky Schupbach works at the 'Q' Bot box at the Massachusetts Institute of Technology's Whitehead Institute in May 2000. The research being done here was part of the Human Genome Project's effort to map all human genes. The project is currently led by Francis Collins.

Collins was born in Staunton, Virginia. His mother taught him at home until he was nine. He studied chemistry at the University of Virginia and earned a Ph.D. in physical chemistry from Yale University in New Haven, Connecticut, in 1974. That same year, he began to study medicine at the University of North Carolina. He became a doctor in 1977.

In 1984, Collins began teaching and doing research at the University of Michigan. In 1989, Collins and his coworkers identified and located the gene that causes cystic fibrosis (**siss**-tik fye-**broh**-siss). *Cystic fibrosis* is a childhood disease that causes certain glands to produce a thick, sticky mucus. The mucus clogs the lungs and blocks glands in the *pancreas,* a digestive organ behind the stomach. Mucus that blocks the pancreas prevents enzymes from reaching the intestines, where they break down food.

The gene that causes cystic fibrosis is a faulty, or *defective,* gene. A person who develops cystic fibrosis—or any of a number of other genetic diseases—must have inherited a defective copy of the cystic fibrosis gene from each parent. A person who inherits just one defective gene is considered a *carrier.* Each time two carriers have a child, there is a 25 percent chance that the child will have cystic fibrosis.

Collins's work in identifying the defective cystic fibrosis gene opened the doors to many other discoveries in cystic fibrosis research. In 1990, for example, scientists successfully made copies of the gene that causes cystic fibrosis when it is defective. The gene that was copied was normal. Then scientists added the gene copies to cells in lab dishes that contained the defective gene. The normal gene corrected the defective genes, preventing the cells from making the thick, sticky mucus. This research may one day lead to more effective treatments for cystic fibrosis—possibly even a cure.

In 1990, Collins led a team that identified the gene that causes *neurofibromatosis* (*nur*-oh-fye-*broh*-muh-**toh**-siss), a set of hereditary disorders that cause soft tumors to develop all over the body. Three years later, the team also found the gene for *Huntington's disease,* a disease of the nervous system that causes uncontrollable jerky movements.

Most recently, Collins has been trying to put together a kind of catalog of tiny variations in the human genome. These variations are known as *single nucleotide polymorphisms* (*pahl*-ee-**morf**-*iz*-umz), or SNPs (commonly pronounced "snips"). By analyzing SNPs, scientists can quickly track down genes that, taken together, can cause a disease to develop. Diabetes, hypertension, and certain mental illnesses are among the conditions caused by a collection of genes.

Collins continues to operate his own laboratory, usually riding a motorcycle to work. He also spends time at a missionary hospital in the country of Nigeria in Africa, where his daughter Margaret is a missionary physician. During one visit several years ago, he saved a man's life by performing emergency surgery with only a handful of instruments. He says that his trips to Nigeria bring him closer to his daughter. Says Collins about his decision to visit his daughter, "It seemed like it would be a wonderful thing to do with your kid."

Francis Crick (1916–)

Francis Crick is famous for discovering the spiral, ladderlike shape of DNA along with James Watson. The shape is often described as a *double helix* (**hee**-liks). Crick and Watson's discovery launched decades of experiments in genetics and is considered one of the most important discoveries in history.

Born in Northampton, England, in 1916, Crick recalls always being interested in science. "I wanted to know what is the world made of," said Crick. "And because I asked so many questions [my parents] bought me something called Children's Encyclopedia...that covered all subjects...the nature of the galaxy and chemistry and how things were made of atoms and so on. And I absorbed this with great enthusiasm and I think I must have at that stage decided to be a scientist."

At age fourteen, Crick entered Mill Hill School in North London, where his education included chemistry, physics, and mathematics. After high school, Crick went to University College London, where he earned a degree in physics in 1937.

After Crick graduated, he stayed at University College London to do research. That research was interrupted by World War II. During the war, he was a scientist for the British Admiralty, working with different kinds of *mines,* weapons that explode when pressure is applied. After leaving the admiralty in 1947, Crick read a book by physicist Erwin Schrödinger (**shruh**-ding-er) (1887–1961) called *What Is Life? The Physical*

This model of a DNA molecule shows its double-helix shape. The molecule's structure was discovered by James Watson and Francis Crick.

Aspects of the Living Cell. Schrödinger wrote that the laws of physics should be applied to biology. Crick found the idea fascinating and switched careers from physics to biology. He was especially interested in *molecular biology* (moh-**lek**-yoo-lur bye-**ahl**-oh-jee), the study of the way that molecules behave within cells.

Crick eventually landed at the Cavendish Laboratory at the University of Cambridge in England. At the time, scientists believed that proteins—not DNA—carried genetic information. Crick wasn't sure that was true. He didn't think that proteins carried genetic material, but he didn't know what did. He became interested in how genes divide and make copies of themselves.

In 1951, a few years after Crick began working at Cavendish Laboratory, James Watson arrived to do research in genetics. Watson and Crick became instant friends. Crick has said that they worked well together because they felt comfortable enough to freely criticize each other's ideas.

Crick and Watson also shared a scientific belief. Oswald Avery (1877–1955) had already shown that DNA carried genetic information. Although many scientists doubted Avery's work, Crick and Watson both thought that it was accurate. They believed that figuring out the structure of DNA would lead to an explanation of how genes copied themselves. The pair received help in figuring out the structure of DNA from the work of Rosalind Franklin (1920–1958), a chemist also working at Cavendish.

Crick and Watson went to work building a model of DNA and gathering whatever information about DNA that they could find—including Franklin's X rays of the DNA molecule. *X rays* are beams of energy that can penetrate most materials, creating images of what's inside. (The images are also called X rays.) Franklin had been using a special kind of X ray to examine the DNA molecule. Her images showed that the DNA molecule's backbone, made up of sugar and phosphate molecules, was located on the outside of the molecule.

The X rays were a key piece of the DNA puzzle ultimately put together by Crick and Watson. They described DNA as being shaped like a double helix, a sort of winding ladder with rungs of chemical bonds between the two sides. Their scientific picture of the DNA molecule—two helical (spiral) chains coiled around each another—formed the basis for the title of their famous book about DNA, *The Double Helix.* Crick, Watson, and a colleague named Maurice Wilkins (1916–) were awarded the Nobel Prize in Physiology or Medicine in 1962 for their work with DNA. Franklin might well have shared the Nobel prize had she not died before it was awarded for this work. Nobel awards are given only to living scientists.

Crick continued his genetic research until 1966, when he switched fields of study once again. This time, he started to work in *embryology,* the study of the cells that make up life in its first few days. In 1977, Crick went to the Salk Institute in La Jolla, California, where he began examining how the brain works. As part of this work, he became interested in the role of dreams.

When he was a boy, Crick once told his mother, "You know by the time I grow up everything would have been discovered." According to Crick, his mother replied, "Don't you worry! When you grow up there will be plenty left for you to discover."

Charles Darwin (1809–1882)

Charles Darwin was born in Shrewsbury, England, in 1809. His father was a wealthy doctor. His grandfather was also a doctor and a well-known nature writer named Erasmus Darwin. His mother's father was Josiah Wedgwood, a well-known businessman who sold the family's valuable Wedgwood china.

Darwin was always interested in nature. As a young man, he went to Edinburgh, Scotland, to study medicine—a family tradition. He soon found that he didn't like medicine. He decided that he wanted to be a minister instead, so he went to Cambridge University in London to learn about *theology* (the study of religion).

Darwin didn't like theology, either, but at Cambridge, he met a botany professor named John Henslow (1796–1861). (*Botany* is the study of plants.) Henslow encouraged Darwin's love of nature. He taught him to be careful and accurate in his observations. He also trained him to collect samples of the plants and animals that he observed. Henslow encouraged Darwin to take a long sea voyage to see more of the world.

Darwin took his advice. In 1831, after graduating from Cambridge, the twenty-two-year-old Darwin signed on with the HMS *Beagle,* a two-masted sailing vessel, as an unpaid naturalist. He spent the next five years exploring the plants and animals of South America.

Darwin was impressed with what he saw in South America. He began to develop theories that were different from the accepted scientific theories of the time. One such theory was that in past ages, Earth had undergone many natural catastrophes, such as earthquakes and volcanic eruptions. These natural disasters wiped out all the animal and plant species alive at that time, and then God created completely new species.

Darwin noticed that some animals he saw were similar to other animals that were considered extinct. An *extinct* species is one that has completely died out. The only evidence that the extinct animals had existed was *fossils,* imprints left behind on natural surfaces such as rocks.

On the Galapagos (guh-**lah**-puh-*gus*) Islands, off the coast of Ecuador, Darwin also observed that each island had its own form of tortoise, mockingbird, and finch. Each of these animals seemed to be related to others of its kind, but each had a slightly different appearance and eating habits. For

This portrait of Charles Darwin, the great naturalist, is from around 1854.

example, Darwin found that the beaks of ground finches ranged in size from large and powerful to small and fine. He thought that the different beaks were related to the birds' feeding habits, not to physical conditions on the islands. Birds with powerful beaks ate large seeds, birds with smaller beaks ate small seeds, and birds with fine beaks mostly ate insects. He thought that each type of finch was matched to the food available in its own environment (en-**vye**-run-ment). (An *environment* is the conditions that surround an organism as it grows, including climate, nutrition, and anything else that affects development.) Darwin believed that the finches had adapted their physical traits to suit their environments. An *adaptation* (*ad*-ap-**tay**-shun) is a change in the body or behavior of a species over many generations, making it better able to survive in its environment.

When he returned to England in 1836, Darwin began filling notebooks with the insights and observations made during his travels. By 1838, he had enough work to publish, but he chose not to. Some experts say that he decided not to publish his findings at that time because he planned a bigger, more impressive work. In 1839, he married and stayed in England to raise his children with his wife, Emma. Because Darwin came from a rich family, he didn't need to earn an income. His wealth let him spend the next twenty years working on natural history projects.

Finally, in 1859, Darwin published *On the Origin of Species by Means of Natural Selection,* which explained his theories of adaptation and a new concept which he called *evolution* (ev-oh-**loo**-shun). Darwin suggested that in animal populations, all animals compete for food. Successful animals live to pass on their genes—and the characteristics carried in those genes—to their children. Darwin called this process *natural selection*. The theory was also referred to as *survival of the fittest*.

Darwin also believed that throughout the years, creatures underwent small, spontaneous changes to adapt to their environments. Some of the changes helped the animals survive. Scientists now call those changes *mutations*. When a mutation gave creatures an advantage in getting food or reproducing, the change was passed on to the offspring, and over time, became one of the common traits of the species. He also believed that all similar species came from a common ancestor, but that the original species had changed in different ways to adapt to different environments. This concept became known as the *theory of evolution*.

An iguana sunbathes in the Galapagos Islands. The many unique creatures in the Galapagos, like the iguana, inspired Charles Darwin's theory of evolution.

Darwin's book sold out on the first day of publication and became an instant best-seller. It is sometimes called "the book that shook the world." Some scientists attacked its theories, arguing that because Darwin couldn't prove his theories, they were useless. Some religious leaders also attacked the book, because Darwin's work seemed to disagree with the story of creation as told in the Bible.

Although he became an instant celebrity after his book was published, Darwin was a shy, reserved man who didn't often attend the public debates that surrounded his work. During the next twelve years, Darwin wrote fourteen books and four papers on the biology of barnacles. He was the world's leading authority on these sea creatures. He wrote about his voyage on the HMS *Beagle*, as well. During that time, he also continued to develop and expand the ideas he had set down in *On the Origin of Species.*

In that book, Darwin had written only of animals and hadn't mentioned humans at all. In 1871, he published *The Descent of Man and Selection in Relation to Sex*. This book said that humans were no different from all other forms of life and that they also had been changed by natural selection. This theory outraged people who believed that his work challenged the Bible.

The next year, Darwin published *The Expression of the Emotions in Man and Animals,* in which he claimed that even our most human behavior—the expression of emotions—reflected our evolutionary past. Once again, religious leaders accused Darwin of being an enemy of religious belief. These religion-based challenges to Darwin's ideas continue to this day, though modern science has proved that Darwin's theories are correct.

Historians believe now that Darwin suffered from *Chagas'* (**shah**-gus) *disease,* an intestinal disease caused by a microscopic organism, *Trypanosoma cruzi* (*trip*-uh-nuh-**soh**-muh **croo**-zye). The organism is spread to humans by insects known in different countries as *vinchuca, barbeiro,* and *chipo.* Their scientific name is *Triatoma* (*trye*-uh-**toh**-muh), and they're known in the U.S. as kissing bugs. Chagas' disease left Darwin tired and suffering from intestinal discomfort for the last years of his life. Darwin died on April 19, 1882, leaving behind his wife, Emma, and eleven children. He was buried in world-famous Westminster Abbey, just a few feet from Sir Isaac Newton (1642–1727), who first identified and explained the law of gravity. Darwin's grave marker reads simply,

CHARLES ROBERT DARWIN
BORN 12 FEBRUARY 1809.
DIED 19 APRIL 1882.

Gertrude Elion (1918–1999)

For almost forty years, Gertrude Elion studied *purines* (**pyoor**-eenz), chemical compounds found in the genes of living creatures. By *synthesizing,* or creating in the laboratory, certain purines, Elion and her coworkers were able to create powerful new drugs that prevented the reproduction of certain viruses and the growth of some cancer cells. Elion's dedicated work led to her sharing the 1988 Nobel Prize in Physiology or Medicine.

Born in 1918 in New York City, Elion was the daughter of Eastern European immigrants. As a child, she had what she later called "an insatiable thirst for knowledge." Her father, a dentist, was financially ruined by the Great Depression, but Elion was able to attend Hunter College, a free university in New York. At Hunter, she majored in chemistry. After seeing her grandfather die of cancer, she hoped to find a cure for the disease. After graduating, she held several jobs, including one as a lab assistant for a chemist. In 1939, Elion entered graduate school. She was the only woman in her graduate chemistry class.

Scientists George Hitchings (left) and Gertrude Elion are shown in this 1988 photo after they won the Nobel Prize in Physiology or Medicine for their work on important medicines to fight bacteria, viruses, and cancer.

Elion earned her master of science degree in 1941. World War II was underway, and corporations had a difficult time finding chemists. This shortage gave Elion a chance at her first real scientific job. She was hired by a food company, but she soon found the work boring, since she was not allowed to do basic research. Her tasks included studying the color of mayonnaise and the amount of acid in pickles. Finally, Elion took a job as a biochemist at the Wellcome Research Laboratories in North Carolina. She worked with George H. Hitchings (1905–1998), who introduced her to purines. Purines are a class of chemicals based on the element nitrogen. Purines are also called *nitrogenous* (nye-**trah**-jen-us) *bases.* These bases are found in DNA.

When Elion began working with Hitchings, scientists did not know the structure of DNA. That information came in the 1950s, with the work of James Watson and Francis Crick (1916–). However, Elion and other scientists did know about the nitrogenous bases and about the role that DNA played in passing traits to offspring. Elion and Hitchings understood that disrupting the activity of the purines within cells could stop the *replication* (rep-lih-**kay**-shun) of DNA, which is the process by which DNA copies itself. Doing this could prevent the growth of cancer cells or kill bacteria and viruses.

Elion's work included synthesizing purines and similar compounds in order to understand how they were created in cells. This work led to the development of drugs that could affect purines in certain cells, halting the cells' growth. The first major drug based on Elion's research was 6-mercaptopurine (mer-*kap*-toh-**pyoor**-een), or 6MP. The drug successfully stopped the spread of *leukemia* (loo-**kee**-mee-uh), a cancer of the blood, in children. In fact, 6MP proved so potent that in 1953, the U.S. government cut short its testing of the drug so that it could be widely sold. Another anti-leukemia drug, 6-thioguanine (*thye*-oh-**gwah**-neen), also came out of this research.

Over the decades, Elion's work led to a number of other helpful medicines. A drug similar to 6MP was given to patients who received organ transplants. The drug helped to stop the body's usual rejection of foreign organs. Other drugs were used to treat infections caused by bacteria, as well as the diseases *gout* and *malaria* (muh-**layr**-ee-uh). One of Elion's later discoveries was that compounds based on sulfur sometimes boosted the effect of purine drugs.

Early in her career at Wellcome Research Laboratories, Elion decided to pursue a Ph.D., the highest college degree that a research scientist can earn. After a few years of study, her school wanted her to leave Wellcome to focus on her studies. Elion, however, decided to give up her education to continue working at Wellcome. Her career never suffered because she lacked a Ph.D. Her success with the purine drugs helped Elion become head of Wellcome's Department of Experimental Therapy in 1967. She also was active in many scientific organizations, including the National Cancer Institute and the World Health Organization (WHO). In 1988, she shared the Nobel prize with Hitchings and British scientist Sir James Black (1924–). Elion was often asked if her goal had been to win the prize. She said no: "What we were aiming at was getting people well, and the satisfaction of that is much greater than any prize you can get."

Elion retired from Wellcome in 1983, but she remained active in scientific organizations. She also served as a consultant, providing advice to her former company. Elion died in 1999, shortly after taking her daily walk near her home in North Carolina.

Alexander Fleming (1881–1955)

Sir Alexander Fleming was proclaimed a Knight of the Royal Court of England by King George VI in 1944 for his discovery of the infection-fighting drug *penicillin* (*pen*-uh-**sil**-in). This discovery changed medical history.

Fleming was born in 1881 in a remote part of Scotland, where his family ran an 800-acre (320-hectare) farm. When Fleming's father died, his oldest brother took over the farm. Another brother was a doctor in London, England. Fleming, several brothers, and a sister eventually went to live in London.

Fleming attended the Regent Street Polytechnic School there. After graduating, he worked for a shipping firm, but he didn't like it. When the Boer War (1899–1902) broke out between the United Kingdom and its colonies in southern Africa, Fleming and two brothers joined a Scottish regiment. The regiment never went to war, but Fleming practiced shooting, swimming, and water polo.

After the war, Fleming decided to study medicine. He chose St. Mary's Hospital Medical School of the University of London because he had played water polo against its team. In 1905, Fleming was studying surgery. However, he changed to bacteriology, because he learned that if he took a job as a surgeon, he would have to leave St. Mary's Hospital. The captain of St. Mary's rifle club wanted to keep Fleming there because he was an excellent marksman. The captain persuaded Fleming to join the Inoculation (in-*ok*-yoo-**lay**-shun) Service. Fleming stayed at St. Mary's for the rest of his career.

Fleming was a playful man who loved games and sports. Fleming is also said to have been rather sloppy. These two characteristics eventually led to his discovery of penicillin. In 1928, Fleming was working to find an *oral* medicine, one that could be taken by mouth, to kill dangerous bacteria. He went on a two-week vacation, leaving a sink full of dishes and some bacteria in a petri dish on his lab bench. A *petri dish* contains nutrients essential for growing bacteria.

When Fleming returned, he found that the bacteria, called *Staphylococcus* (*staf*-ih-loh-**kahk**-us), had grown rapidly in the dish. However, a clear area surrounded a strange, yellow-green mold that had accidentally soiled the plate. All the bacteria around the mold had died. Fleming tested a sample of the mold. He discovered that it contained a chemical that could kill many kinds of bacteria.

The active ingredient in that mold turned out to be a powerful infection-fighting agent. The bacteria it killed included ones that caused scarlet fever and the lung infections *diphtheria* (dif-**theer**-ee-uh) and pneumonia, among many others. Fleming later learned that a spore from a mold called *Penicillium notatum* (*pen*-uh-**sil**-ee-um *noh*-**taht**-um) had drifted through an open window from a fungus lab on the top floor of the hospital. This mold grows on rotting fruits. Fleming named the active ingredient in his mold *penicillin*.

Fleming was not the first to describe the antibacterial properties of *Penicillium notatum*. However, he recognized the importance of the finding. He later said, "My only merit is that I did not neglect the observation and that I pursued the subject as a bacteriologist." In 1929, Fleming published a paper about penicillin in the *British Journal of Experimental Pathology.* Fleming still wasn't sure that penicillin could contribute to medicine. He didn't think that enough of the medicine could be made. For almost ten more years, Fleming's discovery went nearly unnoticed by other scientists.

Fleming stopped working with penicillin, but he kept some of it in his lab. In 1939, two scientists working at Oxford University in England—Howard Florey (1898–1968), an Australian-born physiologist, and Ernst Boris Chain (1906–1979), a German-born chemist—started to work on producing penicillin. After two years, Florey and Chain had made enough pure penicillin to test it on some diseased mice. They proved that injections of penicillin caused amazing recoveries from a variety of infections. During World War II, penicillin was used to cure wound and other infections in soldiers. By the war's end, penicillin was being used by medical professionals all over the world.

Sir Alexander Fleming discovered penicillin in September 1928. The discovery is considered one of the great advances of modern medicine.

Florey, Chain, and Fleming shared the Nobel Prize in Physiology or Medicine in 1945. Because Fleming was the most outgoing of the three, the public accepted him as the sole discoverer of the miracle mold. When Fleming died of a heart attack in 1955, he was mourned by the world and buried as a national hero in the crypt of St. Paul's Cathedral in London.

Rosalind Franklin (1920–1958)

Rosalind Franklin was a female pioneer in the mostly male world of genetic science. Franklin's work was important in discovering the shape of DNA. Franklin was born in London in 1920. She was considered to be sensible and exacting. She decided to become a scientist when she was a teenager.

In 1938, Franklin passed the examination needed to attend Cambridge University in London. Her family was wealthy and had a tradition of public service, but her father didn't think that women should be given a university education. He refused to pay for her education. When Franklin's aunt offered to pay for the education, Franklin's father agreed to let her go.

Franklin graduated from Cambridge in 1941. She then worked at the British Coal Utilization Research Association, researching the structure of coal and carbon. She published five papers about her work before she was twenty-six years old. By the end of her career, she had written seventeen articles on the structure of coal and carbon.

Franklin earned a doctoral degree in 1945. In 1947, she went to the Laboratoire Central des Services Chimiques de l'Etat (lah-*bor*-ih-**twah** sen-**tral** deh ser-**veese** shih-**meek** deh luh-**tah**) in Paris, France. There, she learned how to use X-ray techniques to study the structure of carbon.

Franklin later joined a team at King's College London. The team was studying living cells. Franklin was asked to work on a molecule called DNA. Initially, she believed that she was the only person working on DNA. She soon learned that a colleague, Maurice Wilkins, was also working on the project. The two scientists didn't get along. Franklin was as direct and strong-minded as Wilkins was shy and quiet.

At King's College, Franklin began her groundbreaking study of the structure of DNA. She invented a technique that allowed her to photograph the DNA molecule. The results clearly showed DNA's now-famous spiral, or *helical,* structure. At the time, no one else had been able to produce such photographs.

Rosalind Franklin's X-ray photos of the DNA molecule helped James Watson and Francis Crick discover the double-helix shape of DNA, shown in this model here.

Wilkins shared Franklin's findings—without her permission—with James Watson and Francis Crick at Cambridge University. In March 1953, they published a paper on the structure of DNA in the science journal *Nature*. Franklin didn't know that they had used her research.

When Watson later wrote about the discovery of DNA's structure, he wrote about the first time that he saw one of Franklin's photographs of the molecule. The photo provided exactly the information that he and Crick had needed to fill in the gaps in their theories. Watson wrote, "The instant I saw the picture my mouth fell open and my pulse began to race.... The black cross of reflections which dominated the picture could arise only from a helical structure...."

Franklin eventually wrote several articles about her research into the structure of DNA, but she wasn't happy at King's College. At the time, female scientists were looked down on. In fact, they weren't allowed to eat lunch in the same room as male scientists. Franklin finally went to Birkbeck College in London to head her own research group. There, Franklin finished her work with DNA and returned to her work on the structure of coal. She later studied plant viruses. Her work laid the foundation for the study of *structural virology,* the structure of viruses.

Franklin also conducted research on *poliovirus,* the cause of the muscular disease *polio*. Her research carried great risk, because she was working with a living virus that was able to infect her and the members of her team. After Franklin's death, work on live poliovirus was halted because of that risk.

In 1962, Wilkins, Watson, and Crick received the Nobel Prize in Chemistry for their work on DNA. The award is given only to living people. No one knows whether Franklin also would have shared the award, for she died of ovarian cancer in 1958 at the age of thirty-seven.

Walter Gilbert (1932–)

Walter Gilbert's greatest achievement in biotechnology came because he questioned why different cells produce different proteins even though their DNA is the same. His research into this question led to the discovery of a *repressor gene* in bacteria. A repressor gene turns off certain functions of a cell. Even though cells may have the same DNA, repressor genes make the cells function differently.

Walter Gilbert conducts a lecture at a 1986 meeting held by the Cold Spring Harbor Laboratory in New York. Gilbert's work with genes and hormones earned him the 1980 Nobel Prize in Chemistry.

Gilbert was born in Boston, Massachusetts, in 1932. Gilbert and his sister were taught at home for several years by their mother, a child psychologist. Gilbert's father was an economist who taught at Harvard University in Cambridge. When Gilbert was seven, his family moved to Washington, D.C.

Gilbert was interested in science even as a boy. While still in elementary school, he made mirrors to make his own telescopes. He also joined a group of adults interested in *geology,* the study of the structure of Earth. At age twelve, he conducted his first chemical experiment. It caused an explosion that slashed his wrist and required him to be rushed to the hospital. His mother remembers that despite his pain, he said, "I know what I did wrong!"

Gilbert was bored by school in general but never by science. He was accepted at Harvard University in 1950, where he studied chemistry and physics. After graduating, Gilbert went to Cambridge University in England, where he studied physics. There, he met James Watson, who at that time was already well-known for his work with DNA.

Gilbert returned to Harvard in 1957 as a professor of physics. He later became interested in biology. His interests led him to begin searching for a chemical messenger that carried information from DNA to various areas of a cell. Scientists knew that something must carry DNA's information to the protein factories in cells, but no one knew what that something was.

A year later, Gilbert published his first article on *messenger ribonucleic* (*rye*-boh-noo-**klay**-ik) *acid,* or mRNA. mRNA is like a photocopy of the information carried by a particular gene. A molecule of mRNA carries the gene's information to the parts of the cell that make a particular protein. Gilbert's identification of mRNA added a key piece of the puzzle about how DNA works.

In 1964, Gilbert became a professor of molecular biology at Harvard. It was while teaching and researching there that Gilbert discovered a repressor gene in the common bacteria *E. coli.*

Gilbert's research made him famous among scientists around the world. It also attracted the brightest graduate students to his laboratory. Gilbert made headlines in 1978 when he led a team that discovered how to manufacture insulin from bacteria. People with diabetes usually need to inject themselves with insulin to survive.

Gilbert's discovery opened the door for other bioengineers to use bacteria to make valuable hormones. In 1979, Gilbert created a medicine called *interferon* that fights viral infections. The drug is also used to treat some cancers, as well as *multiple sclerosis* (skluh-**roh**-siss), a nervous and muscular disease that destroys the outer coverings of nerves.

Gilbert was awarded the Nobel Prize in Chemistry in 1980 for his work with DNA. He shared the prize with Paul Berg (1926–) and Frederick Sanger (1918–). In 1982, Gilbert left Harvard to run Biogen, a Swiss-based biotechnology company that he had helped to create. The company, which later did very well, was initially unsuccessful in reaching its goals, and Gilbert resigned in 1984. He returned to Harvard to do research, where he remains today.

William Harvey (1578–1657)

William Harvey was a physician who worked in England around the time of Shakespeare (1564–1616). He was a member of the wealthy upper class and married to the daughter of the physician of Queen Elizabeth I. Later, he became the doctor of King Charles I.

Harvey is best known for his studies of the body, especially the heart. He is also important in biotechnology because he was the first to take a scientific approach to understanding the way *mammals* reproduce. Mammals are animals whose body temperature doesn't change much with their environment. Mammals also have hair and nourish their young with milk. Because humans are mammals, Harvey's work also added to the world's knowledge of human childbearing. He is considered a pioneer in the study of human genetics.

The oldest of seven children, Harvey was born in Kent, England, in 1578. From the start, he was a bright, hardworking student. He earned his bachelor's degree from Cambridge University in England in 1597. He then received a degree in medicine in Italy, at the University of Padua, the leading medical school of the day.

In 1628, in a groundbreaking book titled *On the Motion of the Heart and Blood in Animals,* Harvey described the heart as a pump that moved, or *circulated,* blood. Many scientists and physicians at the time thought that the liver changed food into blood, which was then used as fuel by the body. Others thought that the heart made blood. Harvey's discovery that the heart acted as a pump proved those beliefs incorrect and made him famous throughout Europe.

Harvey observed the development of a chick embryo and was convinced that all animals come from eggs. He developed a theory that female mammals, including humans, had eggs inside their bodies. He searched for a mammal egg by cutting open, or *dissecting,* many animals—particularly deer from the royal deer park, part of a large estate outside York, England. Harvey never found the egg that he was looking for. It wasn't until 1827, more than 200 years later, that scientists found the internal egg of a mammal. Harvey was so well respected that his theory of the internal egg was still widely accepted, even though it could not be proven.

Despite Harvey's explanation of circulation, doctors continued using the technique of *bloodletting* as an accepted treatment for a variety of ailments at the time. In bloodletting, the patient's arm was cut at the elbow, and blood was allowed to drain freely. Bloodletting, or *phlebotomy* (fluh-**baht**-uh-mee), was a common practice from before the time of Hippocrates (hip-**pahk**-ruh-*teez*) (c.460–c.377 B.C.E.), an ancient physician.

Some experts believe that the practice might have started when ancient peoples saw bats removing blood from animals and hippopotamuses scratching themselves on trees until they bled. Ancient doctors might have believed that if bleeding was helpful in some way to animals, it might also be helpful to humans. (The term *phlebotomy* is used today to describe the removal of blood from a patient for blood tests.)

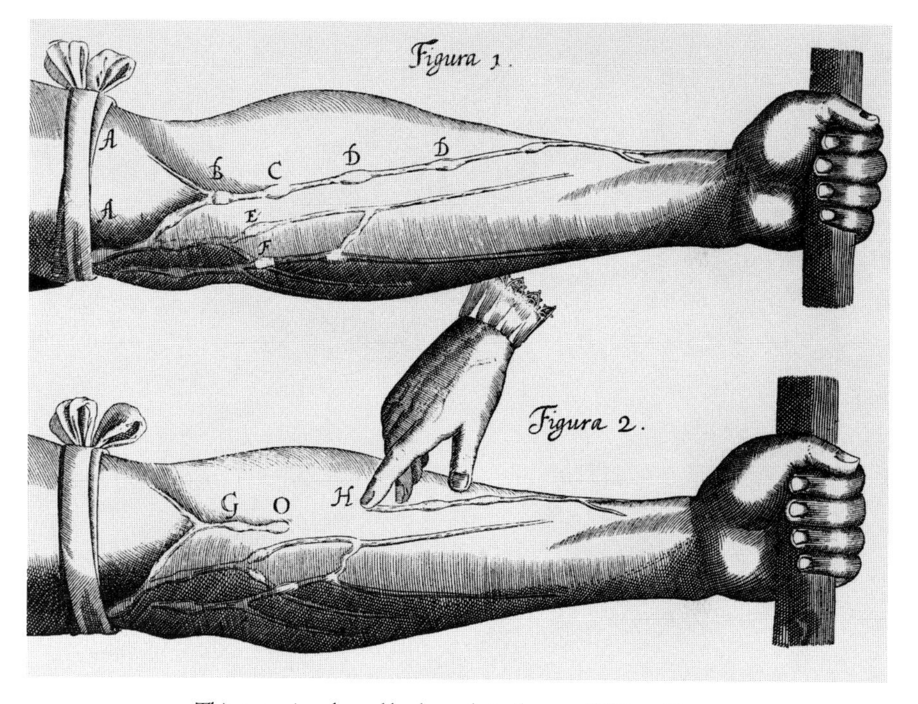

This engraving shows blood vessels in the arm. William Harvey did groundbreaking work with the blood and heart.

Physicians developed a number of explanations to support the use of bloodletting. For instance, for a time, people believed that the body contained four kinds of fluids, or *humors*. The humors were blood, black bile, yellow bile, and *phlegm* (flem)—mucus in the lungs. If a doctor believed that a patient suffered from an imbalance in these humors, a bloodletting was in order. Vomiting was also seen as a cure for humor imbalances. Bloodletting was used to treat just about everything from

William Harvey was a famous physician in the late sixteenth and early seventeenth centuries.

mental illness to fever. It continued to be a popular practice until the twentieth century.

Harvey's discovery of blood circulation was key to a fuller understanding of the body. His work stimulated other scientists to learn more about blood circulation, such as how blood carried air from the lungs. His work with animal reproduction provided a foundation for today's genetic research.

Alfred Day Hershey (1908–1997)

Born in 1908 in the small city of Owosso, Michigan, Alfred Day Hershey was most famous for using an ordinary blender to make a key discovery about DNA. Hershey attended Michigan State University, earning a degree in chemistry in 1930 and a doctoral degree in bacteriology in 1934.

Hershey became a research assistant in the Department of Bacteriology at Washington University, in St. Louis, Missouri, in 1934. He was made an associate professor there in 1942. While at Washington University, Hershey performed research on a type of organism called a *bacteriophage* (bak-**teer**-ee-uh-*fayj*), a virus that infects bacteria. Max Delbrück (1906–1981), a German scientist at Vanderbilt University in Nashville, Tennessee, read Hershey's published articles about bacteriophages. Delbrück and a colleague, Salvador Luria (1912–1991), a physician at the New York College of Physicians and Surgeons, were also working with bacteriophages. Delbrück invited Hershey to Nashville in 1943 to work together on bacteriophage experiments. Hershey accepted, and together, the three scientists organized what they called the Phage Group. Eventually, bacteriophage researchers from around the nation joined the group. They met regularly during the 1940s at Cold Spring Harbor, a famous laboratory center in Cold Spring Harbor, New York.

Hershey and Delbrück noticed that when two different types of bacteriophages infected the same bacteria, the viruses exchanged genetic information. This was the first known example of genetic recombination in viruses. The term *genetic recombination* refers to the joining of genes from different kinds of organisms to give offspring new traits. Hershey and Delbrück showed that when bacteriophages exchanged their genetic information, their offspring behaved differently.

Alfred Hershey is shown at Cold Spring Harbor after it was announced that he had won the 1969 Nobel Prize in Physiology or Medicine.

In 1952, Hershey and American scientist Martha Chase (1930–) performed an experiment that came to be known as "the blender experiment." Hershey and Chase put a mixture of bacteriophages attached to bacteria in a common kitchen blender. They were trying to find a way to tear the bacteriophages away from the bacteria without damaging the bacteria.

A bacteriophage, which looks a little like a lunar lander, attacks a bacterium by landing on its surface. It then bores into the bacterium and injects its own genetic material. Until the blender experiment, scientists thought that bacteriophages injected proteins that carried genetic information. Hershey and Chase, however, found that DNA carried the information, not proteins. Their experiment proved that genes are made of DNA.

Hershey joined the staff at Cold Spring Harbor in 1962. Seven years later, Hershey, Luria, and Delbrück received the Nobel Prize in Physiology or Medicine for their work on the genetic structure of viruses. Their research laid the foundation for a field now known as molecular biology and was also important for the development of more effective treatments for polio, measles, and mumps—diseases caused by viruses. Hershey retired in 1972 and died on May 22, 1997.

Edward Jenner (1749–1823)

Edward Jenner was born on May 17, 1749, in Berkeley, England. He was a surgeon's assistant for nine years before heading to London to study medicine. He returned to his hometown, where he worked as a physician for the rest of his life. Jenner's greatest contribution to biotechnology was his discovery of a medicine that could prevent smallpox. *Smallpox* was a deadly disease that no longer exists.

Edward Jenner is famous for creating a vaccine for smallpox.

Caused by the smallpox virus, the disease is *infectious,* meaning that it spreads easily from person to person. Smallpox, which can be deadly, causes very high fever, aches, and a rash that can leave permanent scars on survivors. For thousands of years, smallpox was one of the most widespread and deadly diseases on Earth. During the twelfth century B.C.E., there was an outbreak of smallpox in Egypt. Even kings weren't safe from the virus. The face of the mummy of pharaoh Ramses V shows evidence of smallpox scars.

In the 1400s, a particularly deadly form of smallpox ran through Europe, killing thousands. The disease struck Mexico in about 1518, killing half the native population in a single epidemic. By the 1800s, smallpox was responsible for the deaths of as many as 3,000 people a year in London.

As a medical student in London, Jenner had seen many people die of smallpox. At the time, many healers tried to prevent smallpox by deliberately injecting, or *inoculating,* healthy people with infected fluid from a smallpox sore. The fluid, called *pus,* was taken from someone with a mild case of the disease. Many people died after being inoculated. When Jenner tried to inoculate some of the people in his town, they refused. They believed that they wouldn't develop smallpox because they had earlier been infected with cowpox. *Cowpox* is a disease that appears most often in cows. Symptoms are similar to those of smallpox, but the disease is not deadly to humans.

Jenner wondered whether the townspeople were right. To find out, he observed local milkmaids, many of whom had developed cowpox from the cows that they milked. He came to believe that cowpox did indeed seem to prevent smallpox. Still, he needed more solid evidence.

Jenner wanted to test his theory that exposing someone to cowpox could prevent the person from becoming infected with smallpox. In 1796, he tested this idea on an eight-year-old boy named James Phipps. He scratched the boy's arm and painted the scratch with cowpox pus. Two months later, he did exactly the same with pus from a smallpox sore. The boy caught cowpox but not smallpox. Phipps survived until the age of twenty, when he died of the lung infection *tuberculosis* (too-*ber*-kyoo-**loh**-siss).

Jenner repeated this experiment twenty-three different times, each time with the same result. Finally, he decided that people who had been infected by cowpox were *immune* (im-**myoon**) to smallpox, meaning they could not get the disease. He called this new method of preventing smallpox *vaccination* (*vak*-sih-**nay**-shun), from the Latin word *vacca,* meaning "cow." In 1798, Jenner wrote about his work on vaccination. Many doctors doubted that his method worked, but when members of the royal family asked to be vaccinated, Jenner's work started to be accepted in England and worldwide.

Vaccinations against smallpox became widespread in England. In 1799, doctors in America accepted Jenner's findings and also began vaccinating people against smallpox. Between 1802 and 1806, the British government gave Jenner funds to continue his studies on vaccination. Vaccination became free for all infants in 1840 and required in Britain in 1853.

Jenner believed that vaccination could some day destroy smallpox forever. Although he didn't live to see it, his dream came true in 1980, when the WHO declared that smallpox had been wiped out throughout the world. The virus itself exists now only in a few laboratories in Russia and the United States.

Jenner suffered a stroke and died early in the morning on January 26, 1823. He was seventy-three years old.

Percy Julian (1899–1975)

Percy Lavon Julian was a true pioneer in biotechnology. His research led him to create many useful products from living substances. He spent a lifetime studying soybeans, creating such products as fire-prevention foam and artificial hormones.

Dr. Percy Julian receives one of his many awards.

Julian's most famous contribution to the field of biotechnology was his invention of a method for making the powerful drug *cortisone* (**kort**-ih-zohn) from soybeans. Cortisone is a hormone used to treat many diseases, including *arthritis* (arth-**rye**-tiss), a painful joint disease; *acne,* a common skin disease; and *systemic lupus erythematosus* (**loo**-pus *er*-uh-*theem*-uh-**toh**-siss), a disease of the body's infection-fighting, or immune, system. Until Julian came along, cortisone was taken from the glands of oxen, which are farm animals related to cows. As a result, the drug was in short supply and expensive to produce. A single drop might cost from $50 to $100. Julian's method made cortisone inexpensive and widely available. The idea of making artificial, or *synthetic,* drugs from plants was revolutionary at the time.

Julian's scientific contributions are even more impressive when considered against the experiences of many black Americans at the time. Although he experienced much unfairness because of his color, he never let it stop him.

Julian was born in Montgomery, Alabama, in 1899. He was the grandson of a former slave. One of six children, he was a bright student. At that time, the city did not provide public education for black students after eighth grade.

Julian's father, a railway mail clerk, and his mother, a homemaker, valued education and encouraged their children to work hard at their studies. Julian entered DePauw University in Greencastle, Indiana. He had to take several courses to catch up on what his public education had not provided. Julian worked his way through DePauw by digging ditches and waiting on tables. He later graduated first in his class with Phi Beta Kappa honors, awarded to only the brightest students.

Julian hoped to earn a doctoral degree to become a research scientist, but racial prejudice blocked his entry into many university graduate programs. At the time, white Americans made it difficult, if not impossible, for black Americans to obtain an education or a well-paying job. This was especially true in the South, where blacks were treated disrespectfully, and in some cases, inhumanely.

For instance, blacks weren't allowed to use the same drinking fountains as those used by whites. When riding a bus in the South, blacks were forced to sit in the back, while whites rode in the front. A number of blacks were also killed solely for being black.

Despite this intense atmosphere of racism, Julian didn't give up his ambition. He became a chemistry instructor at Fisk University, a college for blacks in Nashville, Tennessee. After teaching at Fisk for two years, he received a fellowship in chemistry from Harvard University in Massachusetts. A *fellowship* is an award of money to attend a school. With the fellowship money, Julian was able to complete his master's degree.

After earning his degree with the highest grades in his program, Julian taught at different universities until 1929. During that time, he worked on his doctoral degree in chemistry at the University of Vienna in Austria. At the time, European universities were more open to black students than American universities were.

It was in Vienna that Julian began studying soybeans. He was trying to find out if they contained proteins that might be used to develop useful products. He also became interested in other plant products, including the African calabar bean, a poisonous plant. In 1931, he returned to the United States to continue his research and to head the chemistry department at Howard University in Washington, D.C.

In 1935, Julian again went to DePauw University, where he became the first person to create a chemical called *physostigmine* (*fye*-zoh-**stigg**-meen) from soybeans. Physostigmine is a drug that blocks the action of a chemical named *acetylcholine* (uh-*see*-tul-**koh**-leen). Acetylcholine allows nerves to pass signals throughout the body. It also plays a role in memory. Doctors use physostigmine to block the effect of acetylcholine in patients with such conditions as short-term memory loss, poisoning from carbon monoxide (a colorless gas), and glaucoma (an eye disorder that can cause blindness).

Julian's development of synthetic physostigmine brought him international scientific praise but no professorship. Once again, he was denied a position that he deserved because of his race. DePauw officials refused to make him the head of the chemistry program, though his colleagues wanted him to have the position.

Julian became discouraged with the academic world. In 1936, he wrote to officials at Glidden, a paint and ink company in Chicago, Illinois, asking for samples of soybean oil, often used in inks. W.J. O'Brien, vice president of Glidden, had heard of Julian from business colleagues from the Institute of Paper Chemistry in Appleton, Wisconsin, who had wanted to hire Julian. However, at the time, blacks weren't allowed to stay overnight in Appleton. O'Brien decided to offer Julian a job. Said O'Brien, "If he is half as good as they say he is, I can use him at Glidden."

O'Brien made Julian director of research of the Glidden Soya Products Division. It was the first time that an African-American had been appointed as the director of a major laboratory. Julian remained there until 1954, when he formed his own company, Julian Laboratories.

Julian was able to continue his studies of soybeans while at Glidden. In 1939, he discovered a way to use soybeans to create the hormones *progesterone* (proh-**jess**-ter-*ohn*) and *testosterone* (tess-**tahst**-er-*ohn*). Both chemicals are naturally occurring hormones that play a key role in reproduction. At the time, doctors used progesterone to prevent miscarriages in pregnancy. They also used it with testosterone to treat certain cancers.

During World War II, Julian found a way to make a fire-fighting foam from soy protein. The foam could put out both oil and gas fires. Firefighting personnel in the U.S. Navy called the foam "bean soup" and used it on board ships to put out fires. In 1947, the National Association for the Advancement of Colored People (NAACP) presented Julian with a major award for his achievements.

In 1948, a few years after the war ended, scientists at the Mayo Clinic in Rochester, Minnesota, announced their discovery of cortisone. Julian began working on a way to make artificial cortisone from soybeans. In 1949, his research team created a synthetic cortisone substitute that was less expensive but worked just as well as natural cortisone. Julian's synthetic cortisone cost only pennies per ounce.

During his lifetime, Julian held more than 100 chemical patents. Julian also wrote many papers on his work and received dozens of awards and honorary degrees.

Percy Julian used soybeans such as these to make important advances in biotechnology.

In 1951, Julian and his family moved to Oak Park, Illinois. They were the first black family to live there. His house was firebombed twice by racists. However, most of the community welcomed the Julian family. Today, Oak Park celebrates his birthday as a holiday.

Julian died on April 19, 1975. At his wake, a close friend, historian John Hope Franklin, remembered Julian not just for the contributions that he made to science, but for the kind person he was. "He very much cherished the company of others," said Franklin, "and others cherished his company even more, if such was possible."

Mary-Claire King (1946–)

Mary-Claire King was one of the first geneticists to find one of the genes that cause breast cancer. King was born outside of Chicago, Illinois, in 1946. As a girl, she enjoyed solving problems and puzzles. She was fifteen when her best friend died of cancer, an event that helped shape the course of her future research.

"We had been friends since age seven," King recalled. "Her death was devastating. I didn't know she had cancer: I just knew she was very ill and in terrible pain, and then she died. It seemed so unfair. It wasn't a conscious decision, but I said to myself, something needs to be done. It's the little pebbles that make a path."

King first studied math at Carleton College in Northfield, Minnesota, and later attended graduate school at the University of California at Berkeley to study genetics. During the 1960s, she dropped out of college in protest against the Vietnam War (1964–1975). She eventually returned to her study of genetics.

By 1974, King was searching for an inherited source for some types of breast cancer. She collected blood samples from, and spoke at length with, many families that had members who had been stricken with breast cancer at some point. She was looking for a gene that the cancer victims in each family had in common.

Mary-Claire King's work in genetics has led to advances in the fight against cancer and other diseases. Here, King takes a break during a meeting at the Cold Spring Harbor Laboratory in New York in 1994.

During some of King's research, she discovered that 99 percent of all human genes are identical to chimpanzee genes. Just 1 percent of genetic material separates the two species. King's discovery led scientists to the conclusion that humans and apes probably shared a common ancestor about 5 million years ago.

In 1984, King took a leave of absence from her research. She decided to use her knowledge of genetics to help a group called *Abuelas de Plaza de Mayo,* "Grandmothers of May Square," in Argentina. This group consists of grandparents searching for lost grandchildren who had been kidnapped as babies during a civil uprising in Argentina in the mid-1970s. The babies were taken from their parents and given to military families who wanted children. Most of the parents were murdered. To claim these children and give them back to their original families, the Abuelas de Plaza de Mayo needed a way to prove that the children were biologically related to the people believed to be the real parents. The grandparents also had to prove that the parents raising the children were not related to them.

King traveled to Argentina and worked with the grandparents to identify living grandchildren and reunite the families. She created a DNA test that proved whether a child was related to a set of grandparents. This proof was based on comparing specific genes from the children with the same genes from the grandparents. Many times, King was able to prove beyond doubt that a kidnapped child and his or her grandparents were genetically related. So far, more than fifty children have been found and matched with their families.

When King returned home, she became a professor in the School of Public Health and the Department of Molecular and Cell Biology at the University of California at Berkeley. There, she resumed her work looking for the gene that causes hereditary breast cancer. Her work led to the identification of such a gene, called *BRCA1.*

To find the gene, King searched for a *marker* for breast cancer. Markers are genes located on a strand of DNA close to a related gene. Markers can be easier to find than the related gene. King knew that if she could find the marker for the breast cancer gene, she might be able to match it to the breast cancer gene itself. She and her team searched dozens and dozens of markers before finally finding a match in 1990. The gene itself was discovered by other researchers four years later. King's research team recently reported that the same gene can stop, and in some cases reverse, breast and ovarian cancers.

King's research is now involved with *acquired immune deficiency syndrome,* or AIDS. AIDS is an infectious disease caused by *human immunodeficiency* (im-*myoon*-oh-dih-**fish**-en-see) *virus,* or HIV. The virus attacks the body's immune system. This makes an infected person more likely to develop serious infections or cancers. King is trying to understand why the immune systems of different people infected with HIV respond differently to the virus.

King is now teaching at the University of Washington in Seattle. She tries to teach her students the importance of being persistent and not giving up. "To do science," says King, "you have to not be intimidated by failure, because you're always getting things wrong. Once in a blue moon, everything goes right. Blue moons are rare, but they're very important in science."

Robert Koch (1843–1910)

Robert Koch was a key figure in the study of bacteria and bacterial infections. Koch discovered the bacteria that caused several diseases, including tuberculosis and cholera (**kahl**-er-uh). He also found that fleas on rats spread *bubonic plague,* a bacterial infection that killed thousands of people during the fourteenth century. People at that time called the disease the Black Death.

Robert Koch studied vaccines for many diseases, including tuberculosis, cholera, and bubonic plague. He won the Nobel Prize in Physiology or Medicine in 1905.

Robert Koch was born in Klausthal, Germany, on December 11, 1843. By the time that he was five, he had taught himself to read newspapers. He attended a local high school and became interested in biology. In 1862, Koch went to the University of Göttingen in Germany, to study botany, physics, mathematics, and medicine. He graduated and became a physician in 1866. He worked for a short time at the Hamburg General Hospital and at an institute for retarded children. Then Koch left to open his own practice, caring for people in Langenhagen, Germany.

Koch left his practice to serve his country in the Franco-Prussian War (1870–1871) between France and Prussia. (Prussia, once located in the northern section of Germany, no longer exists.) Following the war, Koch served as a military physician in Wollstein, Germany. There, he did most of his research in bacteriology.

During this time Koch discovered the cause of anthrax. *Anthrax* is an infection that occurs most often in farm animals. Humans can develop the disease from contact with the animals. At the time, scientists didn't know how the disease spread. Koch used his knowledge of bacteria to find out. He used homemade slivers of wood to inject blood from farm animals that had died of anthrax into mice. He injected other mice with blood from healthy farm animals. He found that the mice died when they were injected with blood from diseased animals but not from healthy animals.

Koch next identified the bacterium itself—*Bacillus anthracis* (buh-**sill**-us an-**thray**-siss), a rod-shaped bacterium. His discovery marked the first time that anyone had proved that *microbes* (**mye**-krohbz) or microorganisms, such as bacteria cause disease. Koch also showed how to take bacteria from infected animals and grow them in a laboratory dish.

Koch's research with anthrax led him to outline key rules that he believed should be followed to identify a microbe responsible for a disease. Those rules, called *Koch's postulates,* involve first identifying a specific organism. The organism should then be grown in a laboratory. The lab-produced organism should then be used to cause the disease in experimental animals. Finally, the organism should be recovered from the infected animals. Koch's postulates are still in use today by *epidemiologists* (scientists who study how infections spread) and other medical researchers.

Koch began to research the lung disease tuberculosis in 1881. Tuberculosis was one of the most deadly diseases of the nineteenth century. It usually affects the lungs, but it can involve any part of the body. At the time, no one knew what caused tuberculosis. After Koch discovered the bacterium that caused the disease in 1882, researchers elsewhere began to develop better treatments for it.

Koch traveled to Egypt in 1883 to study cholera during an epidemic there. *Cholera* is an intestinal infection that causes diarrhea. Caused by the bacterium *Vibrio cholerae* (**vih**-bree-oh **kahl**-er-ay), cholera can be mild or serious. In a person with a serious case of cholera, the diarrhea is severe. The loss of bodily fluids can be enormous, leading to dehydration and shock. Unless the person receives treatment, death can come in just hours.

Koch quickly identified the bacterium that causes cholera. He also discovered that the bacteria are spread in water. His work led to a set of rules for handling a cholera outbreak that are still used today.

In 1891, Koch became the director of Berlin's Institute for Infectious Disorders, now called the Robert Koch Institute. He retired in 1904 and was awarded the Nobel Prize in Physiology or Medicine a year later. The award was given in recognition of the work that he had done with tuberculosis. Koch was sixty-six when he died of heart failure on May 27, 1910, in Baden-Baden, Germany.

Anton van Leeuwenhoek (1632–1723)

Anton van Leeuwenhoek loved microscopes and recorded what he saw through them. He never had formal scientific training or went to a university. He built microscopes as a hobby. Leeuwenhoek put together nearly 250 microscopes. Some of his microscopes could magnify objects up to 270 times.

Dutch naturalist Anton van Leeuwenhoek made simple microscopes through which he observed microbes.

Born in Delft, the Netherlands, Leeuwenhoek was the son of a Dutch basket maker. His working life began at the age of sixteen, when he was sent to Amsterdam to become an apprentice to a clothing merchant. Eventually, Leeuwenhoek returned to his hometown of Delft, where he lived for the rest of his life. He continued to work in the clothing business and began making and using microscopes to inspect the quality of his fabrics.

Leeuwenhoek wrote about everything that he saw through his microscopes and made sketches of key details. His goal was to study as many things as possible under his microscopes. Using his shaving razor, he cut thin slices of cork, plants, and other objects. He examined drops of water from a nearby lake and recorded all the various microbes that he found. He noticed that tiny organisms covered almost everything that he observed. He called these organisms *animalcules*.

Leeuwenhoek shared his discoveries with other scientists by writing letters to Europe's leading scientific societies. Over the course of fifty years, he wrote more than 300 letters. For instance, a letter dated September 7, 1674, provided what even today would be an accurate description of a green algae called *Spirogyra*. He described clearly the coiled appearance of the algae under a microscope and even described the size:

> *Passing just lately over this lake...and examining this water next day, I found floating therein...some green streaks, spirally wound serpent-wise, and orderly arranged.... The whole circumference of each of these streaks was about the thickness of a hair of one's head. All consisted of very small green globules joined together: and there were very many small green globules as well.*

In response to Leeuwenhoek's letters, the Royal Society, an organization of scientists in London, sent an observer to Delft in 1673. The observer was so impressed and sent back such flattering reports that Leeuwenhoek soon received visits from royalty, including the queen of England and the czar of Russia. They wanted to see what Leeuwenhoek saw under his microscopes.

Leeuwenhoek was the first scientist to give a complete description of red blood cells. He also was the first to see the tiny organisms now known as bacteria and the tiny, one-celled animals now called *protozoa.*

Leeuwenhoek was elected as a member of the Royal Society in 1680. He continued to make observations and show others what he had seen right up until his death on August 30, 1723. He was ninety-one years old.

Rita Levi-Montalcini (1909–)

Rita Levi-Montalcini, a physician and researcher, discovered *nerve growth factor,* an important body chemical needed for the proper growth of nerve cells. Levi-Montalcini was born in Turin, Italy, on April 22, 1909. Her father, who believed that a university education would interfere with a woman's role as wife and mother, didn't want any of his three daughters to have a professional career. However, when Levi-Montalcini was twenty, she asked her father for permission to pursue a career. At last, he agreed and allowed her to attend high school.

After graduating from high school, Levi-Montalcini enrolled at the University of Turin. She graduated in 1936 with a degree in medicine and surgery. That year was a difficult one for Levi-Montalcini, who was Jewish. Benito Mussolini was ruling Italy at the time. Mussolini was a dictator who would one day join forces with Adolf Hitler, leader of the Nazi party in Germany. The Nazis were *antisemitic,* or prejudiced against Jews. Mussolini's government passed laws forbidding Jewish people from having professional careers. He considered Jews to be second-class citizens and tried to force Italian Jews out of any political offices that they might have held. Levi-Montalcini and her family considered moving to the United States in order to escape Italy's antisemitism, but decided to stay and try to survive the turmoil to come. Levi-Montalcini built a small laboratory in her bedroom in order to do research.

Italian Agriculture Minister Alfonso Pecoraro Scanio speaks during a press conference in Rome, Italy, in February 2001, with Rita Levi-Montalcini. Levi-Montalcini and other leading Italian scientists went before the Italian parliament to urge the government to loosen restrictions on gene research.

Eventually, the family had to leave Turin, as World War II made it too dangerous to live in the city. They faced the possibility of being killed in a bombing raid or being rounded up by Italian police and sent to a German prison camp. There, they almost certainly would have been killed. They moved to a country house in Piemonte, Italy, where Levi-Montalcini built a new lab and went on with her research. In 1943, the German army invaded Italy, and Levi-Montalcini and her family moved to the city of Florence. They lived there in hiding until the end of the war.

Levi-Montalcini returned to Turin and the university in 1945. Two years later, she moved to the United States to teach and perform research at Washington University in St. Louis, Missouri. Levi-Montalcini's research there led to the discovery of nerve growth factor.

Growth factors are proteins that control the growth of different kinds of cells and help repair damaged *tissues*—collections of similar cells that perform the same functions. Levi-Montalcini discovered nerve growth factor almost by accident when she placed tumors from mice into chick embryos. She noticed that the nervous systems of the embryos grew more rapidly than normal. Knowing that the tumors had never touched the nerves, she came to believe that the tumors made a substance that caused the nerves to grow. She soon found that substance, which she called nerve growth factor.

When Stanley Cohen joined the staff at Washington University in 1953, Levi-Montalcini and Cohen performed even more research on growth factors. They later received the Nobel Prize in Physiology or Medicine in 1986 for their work with growth factors. The work of Levi-Montalcini and Cohen laid the foundation for a field of medical research that continues today. In October 2001, for example, researchers at Cornell University in Ithaca, New York, announced that they had used nerve growth factor to help transplant brain cells in adult rats. Results from that research may one day prove helpful in the treatment of injuries of the spinal cord, a thick bundle of nerves running from the base of the brain to the bottom of the spine.

Levi-Montalcini became a U.S. citizen in 1956. She created a research department in Rome, Italy, and spent part of her time in St. Louis and part in Rome. Levi-Montalcini retired from Washington University in 1979 and returned to Italy to live and work.

Salvador E. Luria (1912–1991)

Along with Alfred Hershey (1908–1997) and Max Delbrück, Salvador E. Luria was a pioneer in the study of viruses and how they reproduce. This work led to discoveries about genetic recombination in viruses. Luria was also active in many political and social causes, which sometimes made him unpopular with political leaders in the United States.

Luria was born in Torino, Italy. He studied medicine at the university in his home city, obtaining his medical degree in 1935. Three years later, Luria, who was Jewish, left Italy for France. At the time, the Italian government, like its ally Germany, had established policies that limited the rights of Jews. Luria spent two years in France studying *radiology,* the use of radiation in science, continuing work that he had begun in Italy. (The term *radiation* refers to the release of energy in the form of waves.) Luria was particularly interested in how radiation can cause mutations in the offspring of fruit flies.

Shortly after the start of World War II, Luria left France for the United States. He did research at several universities before going to Indiana University in 1943. By this time, he had already begun to study bacteriophages, viruses that infect bacteria. Because of their simple structure, "phages" were easy to examine in a laboratory. These viruses also reproduced very quickly and in large numbers, so Luria could easily see changes from one generation of bacteriophages to the next. In 1941, Luria worked with Delbrück at the Cold Spring Harbor Laboratory, studying phages. Other scientists joined the effort, and they were collectively called "the Phage Group."

Max Delbrück looks on as Salvador Luria examines a specimen at the Cold Spring Harbor Laboratory in New York in 1941. Their work in genetics, along with that of Alfred Hershey, earned the three men the 1969 Nobel Prize in Physiology or Medicine.

Luria learned that mutations occur randomly in the genes of bacteria. These random mutations were said to be spontaneous. Some mutations helped the bacteria develop resistance to the invading phages. Delbrück then worked out a mathematical formula that would estimate the chances that a given mutation would occur during a specific bacterium's lifetime. This work led scientists to focus on bacteria when doing research on genes and mutations, since the bacteria were easy to work with in a lab.

In 1945, Luria made another discovery. When a bacteriophage showed a spontaneous mutation, the bacteria that it infected showed the same mutation. (Working on his own, Hershey made the same discovery.) Luria believed that the genetic material of the phage somehow mixed with the genes of the bacteria.

Luria, along with Hershey and Delbrück, continued to study the nature of the genes in phages, work that eventually led to the discovery of genetic recombination in viruses. Hershey and Delbrück did the basic research showing that genes from two different viruses can combine and produce offspring with different traits. Luria then conducted experiments that proved the idea. He used radiation to damage the genes of different phages and then placed them in bacteria. The phages' genetic material combined, and the offspring of the phages did not show signs of damage.

The work that Luria, Hershey, and Delbrück did with phages earned them the 1969 Nobel Prize in Physiology or Medicine. By that time, Luria was a respected professor at the Massachusetts Institute of Technology. He was also an outspoken critic of the U.S. role in the Vietnam War. His positions are most likely what led the NIH to put Luria on a list of scientists who could not serve on certain of its review panels.

Luria believed that scientists had a duty to speak out on important issues, especially when science was involved. In 1977, he wrote, "Scientists should actively promote open discussion of the goals and limitations of science in order to generate informed public opinion." He believed that this dialogue was especially important in the field of genetics, where open discussion would ease any fears of the new technology—one that he had helped shape decades before.

Barbara McClintock (1902–1992)

Women in science have faced many obstacles. For years, many people believed that women couldn't understand difficult scientific ideas. This belief led to roadblocks that kept women out of science. Barbara McClintock's significant achievements in genetics are all the more remarkable because they were achieved when women were often blocked from working in science.

McClintock discovered that genes can move from one place to another on a chromosome. Genes that move on a chromosome have been referred to as *transposons,* commonly called *jumping genes*. The discovery of jumping genes helped scientists better understand how bacteria can withstand the effects of infection-fighting drugs. These genes are also involved in the process of a normal cell turning into a cell that can cause cancer.

McClintock was born in Hartford, Connecticut, on June 16, 1902. Her mother, a skilled pianist, taught McClintock how to play the piano. When her family moved to Brooklyn, New York, she attended high school there and developed a keen interest in science. Her parents didn't want her to go to college, because it was not common for women to get a higher education at that time.

McClintock was determined to continue her education, however. She took a job and found a tuition-free college—the Cornell University School of Agriculture. She entered Cornell in 1919 and studied biology. She was fascinated by genetics. Hoping to continue studying the topic in graduate school at Cornell, she asked for permission from the university to study genetics in the field. Her request was denied because she was a woman. As a graduate student, therefore, she joined the botany department and studied plants instead.

During her studies of botany, McClintock began to look at corn cells. She focused her attention on colorful corn. The kernels on one ear of such corn are typically colored differently from kernels on other ears. McClintock studied the corn cells and began to wonder why the multicolored patterns on ears of the corn didn't follow the usual rules of genetics. She learned to identify corn chromosomes and link each one to the trait that it controlled. Her work at Cornell formed the basis for the discovery that she would eventually make.

McClintock received her doctoral degree from Cornell in 1927. She began her scientific career as a researcher at Cornell and then moved to the University of Missouri. She continued studying corn and pioneered the field of corn *cytogenetics,* the study and analysis of genetic information in corn cells.

McClintock began to see the connection between certain traits that passed from generation to generation and the location of the genes for such traits on the chromosomes. She also recorded the effects of X rays on corn cells.

McClintock loved her work and said that she sometimes felt that she herself was part of the cells that she studied. She liked to tell students to "listen to the plant" for guidance on how to proceed with an experiment. She was also known as an instructor who didn't pamper her students. A colleague once said that McClintock's motto was "Let them sink or swim!"

McClintock's reputation as a brilliant, independent-minded scientist didn't earn respect from many of her male colleagues, who didn't like being outperformed by a woman. Finally, in 1941, she left the University of Missouri to work at the Carnegie Institute at Cold Spring Harbor, near New York City. There, for the next forty-three years, McClintock studied the

Dr. Barbara McClintock (right), and Dr. Louis Sokoloff pose with their Albert Lasker Awards in New York City in November 1981. The Lasker Award is the most prestigious American prize for medical research.

changes in color and texture of corn kernels and the leaves of growing plants. It was there that she realized that certain genes seemed to move from cell to cell as the corn kernel developed. She later explained, "[One] cell gained what the other cell lost."

McClintock published a paper about her findings in 1950. She expected other scientists to realize the importance of her discoveries, but instead, her paper was met largely with silence. For years, many of McClintock's research findings went unrecognized by other scientists. By the 1960s, however, scientists studying bacteria had begun to see the same evidence of jumping genes that McClintock had discovered ten years before. The new information proved that she had been correct about jumping genes.

In 1981, McClintock was awarded the Nobel Prize for Physiology or Medicine for the work that she had done decades earlier. Yet even the Nobel didn't earn her the respect that her male colleagues enjoyed for similar achievements. Once, after receiving the Nobel, she was invited to a dinner by Henry Kissinger, who had served as secretary of state for President Richard Nixon. Kissinger had himself won a Nobel Peace Prize in 1973. Many other leading scientists who had also won Nobel prizes attended the dinner. McClintock noticed that the invitations for her male colleagues addressed them as "Dr.," in recognition of their having earned a doctorate degree. McClintock's invitation read simply "Ms.," despite her own doctorate. She is said to have written on the invitation a famous quote from comedian Rodney Dangerfield: "I don't get no respect!"

McClintock died in 1992 at the age of ninety. Howard Green (1925–), a cellular biologist at Harvard University in Massachusetts was a colleague and a dear friend of McClintock's. After she died, Green wrote of the impact that McClintock had on his life:

I can say that knowing Barbara has been one of the great experiences of my life and the fact that she is gone makes me think of...a miraculous creation that will never again be seen in the world.... If she had made no important discoveries, I would feel about her almost as I do now. Those of us who knew her will...marvel at what genes and experience gave to her alone.

Gregor Mendel (1823–1884)

Gregor Mendel's short scientific paper, "Experiments with Plant Hybrids," described how traits are passed from generation to generation. It has become one of the most important publications in the history of science. However, when Mendel presented the paper to colleagues at a science society in 1865, no one gave it much attention.

More than a hundred years after his death, this Austrian monk is more well known than he was in his own lifetime. Mendel was the second child of Anton and Rosine Mendel, farmers in Brunn, Moravia (a region in the Czech Republic). He was a brilliant student, and his parents wanted him to receive a higher education. They couldn't afford to send their son to college, so Mendel entered an Augustinian monastery (a religious institution operated by Augustinian monks). He continued his education there and began to teach science.

Mendel loved nature. He often took walks around the monastery. During one of these walks, he came across a plant that appeared different from other plants of the same species. He had always been interested in why some plants showed unusual traits and others didn't. He took the strange plant and planted it next to typical plants of the same species.

Mendel wondered if the location of the odd plant had somehow caused the unusual traits. He suspected that if the plant's original location had indeed caused the changes, then the plant's offspring would also show the same unusual traits in a new location. His suspicions proved correct. Mendel realized that traits are passed from generation to generation, or inherited.

Mendel was filled with natural curiosity. He began growing pea plants in the monastery gardens and continued his experiments "for the fun of the thing." He became so attached to his studies that he called the plants his children.

Mendel did his experiments very carefully. For instance, he would delicately dust flowers with the pollen from another plant. Pollen is made up of tiny grains from seed-producing plants. Pollen grains released by the flower of a plant are carried to other plants by wind, insects, birds, or other animals. Pollen from one plant that lands on another plant may pollinate the plant, meaning that the male pollen grains join with female cells in the plant. Pollination results in the production of seeds. After Mendel pollinated a flower with pollen from another flower, he would place a bag over the pollinated flower to make sure that no other pollen would land there. In that way, he could be sure that only the pair of plants that he wished to combine would be the parents of the offspring.

Mendel noticed that some traits were inherited more often than others. He called the more commonly inherited traits *dominant* (**dahm**-ih-nunt). He also noticed that after dropping out of sight for several generations, some traits suddenly reappeared. He called these traits *recessive* (ree-**sess**-iv). Mendel decided that even though recessive traits didn't appear for several generations, they had always been there, waiting to reappear. He began studying how often different traits showed up in the offspring of plants.

After many experiments, Mendel concluded that traits are controlled by what he called *atoms of inheritance,* which never break or blend. Today, these atoms of inheritance are called *genes.*

Gregor Mendel's work with pea plants in the 1800s formed the basis for the field of genetics.

Mendel's work eventually became the foundation for modern genetics. Because of his research, scientists now know that the genes for many diseases are inherited. Scientists also now can use genetic engineering to produce new plants with desirable traits. In fact, many of the most common items available today wouldn't have been possible without Mendel's research. From apples in a produce department to flowers in a greenhouse, the results of the foundation that Mendel laid so long ago are still with us today.

Yet in Mendel's time, other scientists didn't accept his work. After all, Mendel was a monk, not a scientist. Yet Mendel knew that his findings would someday be proved correct, after all. Not long before his death, in Brunn, Germany, on January 6, 1884, he uttered what may be his most famous words: "My time will come."

Lady Mary Wortley Montagu (1689–1762)

Lady Mary Wortley Montagu was a brilliant and high-profile individual in her day. She was a poet, feminist, and world traveler. Lady Mary was often seen at social events. The hundreds of letters that she wrote to friends, including many well-known poets and political figures, were eventually published. These letters made her famous in her day and for almost three centuries afterward.

Lady Mary is best known for introducing the idea of smallpox inoculation to England. The term inoculation refers to deliberately exposing a person to a small amount of a virus so that the person develops a mild case of the disease. After recovering, the person is then immune to the disease. The body's infection-fighting system has learned how to prevent the virus from attacking the body.

Lady Mary was born in England to a rich, upper-class family. Like other upper-class girls, she was educated at home. By the age of eleven, she had taught herself Latin using the books in her father's library. She had also sent a letter to a London bishop in favor of a woman's right to obtain a formal education.

Lady Mary Wortley Montagu, who brought the idea of smallpox inoculation to England in the 1700s, is shown in the clothing that she wore while her husband was ambassador to Turkey.

When Lady Mary was a child, smallpox killed her brother and left her with no eyelashes and a pockmarked face. Smallpox was a deadly disease that no longer exists. Caused by the smallpox virus, the disease is infectious, meaning that it spreads easily from person to person. Smallpox causes fever, aches, and a rash that can leave permanent scars, as it did on Lady Mary's face. For thousands of years, smallpox was one of the most widespread, deadly diseases on Earth. During the twelfth century B.C.E., there was an outbreak of smallpox in Egypt. Even kings weren't safe from the virus. The face of the mummy of pharaoh Ramses V shows evidence of smallpox scars. In the 1400s, a particularly deadly form of smallpox ran throughout Europe, killing thousands. The disease struck Mexico in about 1518, killing half the population there. By the 1800s, smallpox was responsible for the deaths of as many as 3,000 people a year in London.

In 1716, at the age of twenty-seven, Lady Mary moved to Constantinople (now called Istanbul) in Turkey with her husband, who had been appointed ambassador to that city. A year later, Lady Mary wrote of her travels in the Middle East. The collection of her letters from this period is now known as "the Turkish Embassy Letters." In these letters, she wrote about the local practice of giving people a mild form of smallpox to protect them against catching a more violent, life-threatening case of the disease.

The process, called *engrafting* in Turkey, involved taking material from a smallpox scab and spreading it on an open cut on the person being inoculated. Lady Mary had her three-year-old son inoculated so that he would not get smallpox, as she and her brother had. When she returned to England, she spoke to many doctors and politicians about the value of inoculation. Because she was a prominent person in England, others took up her cause. Lady Mary later insisted that her English doctor inoculate her five-year-old daughter.

During the 1700s, smallpox was widespread throughout England. Many of the people who knew of Lady Mary's push for inoculation had themselves inoculated. Most were saved from the death that smallpox often caused.

However, inoculation alone couldn't stop smallpox from killing people. Some people who had been inoculated died of the disease anyway, or they lived but carried the virus and spread it to others. It wasn't until Edward Jenner (1749–1823) discovered vaccination that the disease began to be controlled. *Vaccination* is basically a more refined method of inoculation that proved safer and more effective than the inoculation available in Lady Mary's day.

Lady Mary's tireless work at persuading people to become inoculated saved countless lives. She showed England and the world a new way of thinking about smallpox and how it might be cured.

Lady Mary died of breast cancer in 1762. She was seventy-three years old.

Kary B. Mullis (1944–)

Kary Mullis developed the *polymerase* (**pahl**-uh-mur-ayce) *chain reaction,* or PCR, a valuable tool used routinely by today's molecular biologists. PCR is used to make multiple copies of DNA in a short time.

Mullis was born in Lenoir, North Carolina. He attended college at the Georgia Institute of Technology, where he studied chemistry. After graduation, he earned a doctoral degree in biochemistry from the University of California at Berkeley in 1972.

In the late 1970s, Mullis took a job at Cetus Corporation, a biotechnology company in California. His work involved making tiny sections of DNA called *nucleotides.* A nucleotide consists of a *base*—one of four main chemicals in DNA, either adenine (**ad**-uh-*neen*), thymine (**thye**-meen), guanine (**gwah**-neen), or cytosine (**sye**-toh-*seen*)—plus one molecule of sugar and one of phosphoric acid. Nucleotides are the building blocks of DNA. Millions of nucleotides linked together in a coiled chain form a DNA molecule.

In 1983, Mullis began to develop a way to identify particular nucleotides in a section of DNA. While working on this technique, he came up with something very different—PCR. Mullis says he thought of it while driving his car down a California highway. The idea came to him so suddenly that he pulled his car to the side of the road, stopped, and eagerly explained the concept to a friend riding with him.

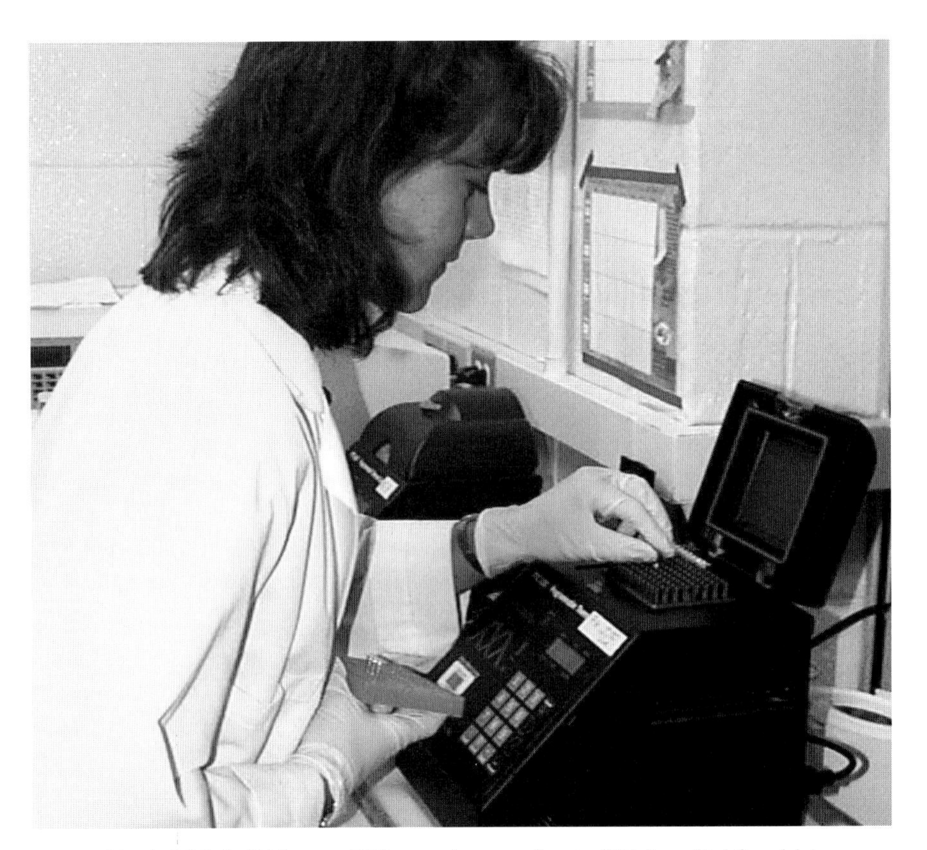

Scientist Michelle Shipley uses PCR on anthrax samples at a U.S. Army biodefense lab in December 2001. U.S. investigators hoped that PCR tests would help them track down the person that sent anthrax-contaminated letters through the U.S. mail system at the time. Dr. Kary Mullis developed PCR in the mid-1980s.

PCR is a technique that can copy, many times over, a small sample of DNA. DNA's structure is like a winding ladder with rungs of chemical bonds between the two sides. Mullis's process involves heating a vial containing a section of DNA until the two strands of the molecule split apart. When the DNA cools again, each strand forms a second strand, identical to the one that has broken off. It is like taking a ladder apart, leaving it overnight, and finding two identical ladders in the morning. Mullis used a chemical called *polymerase* to make the process work.

Each heating-cooling cycle in the PCR process doubles the amount of DNA. Running through the cycle just forty times multiplies the amount of DNA a trillion times! Huge amounts of DNA like this are needed to study the molecule in detail.

PCR is used in a number of ways today. For instance, PCR is a key process for *DNA profiling,* also called *DNA fingerprinting*. DNA profiling is a process that uses a tiny amount of DNA to create a kind of chemical bar code that can be compared with other bar codes. Comparing one person's DNA profile with a DNA sample from a crime scene can help prove the guilt or innocence of an individual accused of a crime. PCR is also a valuable tool for medical research and even agricultural biotechnology.

Mullis published the first description of the PCR process in 1985. After the paper was published, he wanted to receive credit for the invention and be able to control how it was used. However, his employer wanted to control the process. Mullis left Cetus Corporation in 1986. A year later, Cetus obtained a patent for PCR.

Mullis is currently conducting research on HIV. Most scientists believe that this virus causes the disease AIDS. Mullis doesn't think that HIV is the cause of AIDS, though he's not sure what is.

In 1993, Mullis was awarded the Nobel Prize in Chemistry for his development of PCR. Mullis shared the prize with Michael Smith (1932–2000), a biochemist from Vancouver, Canada, who developed a process similar to Mullis's. Mullis is also an author, and he lives in California.

Severo Ochoa (1905–1993)

Severo Ochoa achieved a number of important scientific advances, but he is connected to biotechnology because of his work with *ribonucleic acid,* or RNA. RNA is a large molecule in a cell that carries instructions from DNA molecules to other parts of the cell. In 1955, Ochoa became the first person to create RNA in a lab.

Ochoa was born in Luarca, Spain. At the age of twenty-four, he received a medical degree from the University of Madrid in Spain. He later studied in Glasgow, Scotland, and in Berlin and Heidelberg, Germany. During this period, he began to concentrate his studies on enzymes. Ochoa discovered several enzymes that proved valuable in medical research being done elsewhere.

By 1940, Ochoa believed that there was more opportunity for scientific research in the United States. He emigrated there in 1941 and became a citizen in 1956. Ochoa taught briefly at Washington University in St. Louis, Missouri. In 1942, he joined the faculty of the College of Medicine of New York University, where he worked for thirty years as a professor, researcher, and eventually, head of the biochemistry department.

In the 1950s, Ochoa studied the actions of chemicals during *photosynthesis* (*foh*-toh-**sin**-thuh-siss), a process in plants that uses light to turn water and the gas carbon dioxide into food for the plants. It was during this period that Ochoa discovered a way to make artificial RNA. In 1955, he found that an enzyme could be used to join smaller molecules together to make RNA. This marked the first time that molecules had been joined in a chain outside of a living organism.

Dr. Severo Ochoa is shown in his laboratory at New York University in October 1959, the same year that he won the Nobel Prize in Physiology or Medicine for his work with RNA.

The achievement won Ochoa many awards, including the Nobel prize in 1959. He shared the prize in physiology or medicine with American biochemist Arthur Kornberg (1918–). Ochoa was honored for his research on RNA, and Kornberg was honored for research on DNA. Ochoa died in 1993 from pneumonia.

Louis Pasteur (1822–1895)

Louis Pasteur was one of the greatest scientists of all time. His work with bacteria and other organisms made him an early pioneer in biotechnology.

Pasteur was born in Dôle, France, on December 27, 1822. He was the son of a *tanner,* a person who makes leather out of animal skins. Pasteur grew up in the small town of Arbois, France. In 1847, he earned a doctoral degree at the École Normale, a university in Paris, where he had studied physics and chemistry. Even as a student, he was known for his brilliant scientific mind.

Pasteur spent several years teaching and doing research at colleges in Dijon and Strasbourg in France. In 1854, he became professor of chemistry at the University of Lille. He also became head of a team of professors who were finding ways of applying the principles of science to the practical problems of industries in the region. One of the largest industries there was wine making.

In 1856, Pasteur observed tiny organisms in wine that had soured. When he heated the sour wine, the organisms, which he called *germs,* disappeared. He also noticed that when wine was covered, it stayed free of germs. As soon as the cover was removed, though, the germs returned, and the wine eventually went sour.

Pasteur performed several experiments to find out where the germs came from. For centuries, many people had believed in the theory of *spontaneous generation* (spon-**tay**-nee-us jen-ur-**ay**-shun). They thought that small life forms, such as lice and fleas, were born from dead, decaying matter. Like other scientists before him, Pasteur didn't share that belief.

The great scientist Louis Pasteur, who made many contributions to biotechnology, is shown here in his laboratory study.

Based on his experiments, Pasteur thought that the germs were blowing around freely in the air. With this idea in mind, he invented the process of *pasteurization* (*pass*-chur-ih-**zay**-shun). During pasteurization, a fluid is heated just enough to kill the germs—what are now known as bacteria. After being heated, the fluid is instantly covered to keep the bacteria out. Pasteurization, a process still used today, prevents the fluid from spoiling. The process is used on wine, milk, and other beverages.

This illustration shows Louis Pasteur at work, looking through a microscope.

A number of people, such as Anton van Leeuwenhoek (1632–1723), had already observed bacteria. Leeuwenhoek called the microbes *animalcules,* but he didn't know what role they played in nature. It was Pasteur who finally realized that bacteria caused disease. In 1864, Pasteur stated his belief to other scientists that germs are in the air and that some diseases are passed from person to person through the air. He called this idea the *germ theory*.

By 1865, Pasteur had become director of scientific studies at the École Normale. He was called to Paris to help France's silk industry. The country produces large quantities of silk. The entire industry was in danger because a disease was close to destroying all the silkworms in the nation.

Silkworms are actually moths in the wormlike, or *larval,* stage of development. In certain kinds of moths, the larva produces a thin, long string of silk to make a *cocoon,* a protective shell surrounding the larva. To make the cocoon, openings near the mouth of the larva release a liquid that becomes hard when it comes into contact with air. Silk from these cocoons is used to produce silk clothing and other objects.

Pasteur suspected that germs found in the diseased silkworms were the cause of the disease. Pasteur's experiments eventually discovered which microbe was responsible for causing the disease. The germ survived by destroying the silkworm. Pasteur believed that he could prevent the germ from destroying the silkworms in France by making sure that silk was taken only from healthy larvae that had hatched from germ-free eggs. His experiments saved the silk industry from disaster.

The rest of Pasteur's life was spent developing *vaccines.* A vaccine is a mixture of weakened or killed disease-causing organisms, such as bacteria or viruses. When a vaccine enters a person's body, it affects the person's immune system. The vaccine sparks the immune system to produce special proteins, called *antibodies* (**an**-tih-*bahd*-eez), that protect the person against the organism. The weakened or killed organisms aren't able to cause severe infection. The antibodies produced can give the person protection, or *immunity,* from the disease for months, years, or in some cases, a lifetime.

Pasteur's most famous work with vaccines involved rabies. *Rabies* is a disease caused by the rabies virus, a bullet-shaped virus that causes swelling of the nervous system. In humans who become infected, rabies causes fever, headache, and confusion. The disease is almost always fatal in humans. Most humans who develop rabies get it from the bite of an animal infected with the virus. Bats, wolves, raccoons, skunks, foxes, and coyotes are particularly susceptible to the rabies virus.

Pasteur believed that a vaccine made from weakened rabies virus might prevent people from dying from the disease or even from developing it at all. At the time, the only treatment available for rabies was to immediately burn the bite wound with a red-hot poker. This treatment usually failed because the virus quickly enters the bloodstream. When that happens, the person dies a painful death.

Pasteur and an assistant, Emile Roux (1853–1933), performed several experiments on animals to see if Pasteur's vaccine worked. Each time that he vaccinated a dog and then exposed it to rabies, the dog survived. Although Pasteur insisted that his vaccine wasn't ready to be tested on humans, he found himself, on July 6, 1885, needing to treat a nine-year-old boy who had been bitten by a dog with rabies. Over the course of ten days, Pasteur gave the boy a vaccine made from the strongest rabies virus known. The boy recovered and remained healthy. Since that time, Pasteur's vaccine has saved countless lives.

In 1888, Pasteur's research on rabies resulted in the creation of a special institute in Paris for the treatment of the disease, the Pasteur Institute. Pasteur was the director of the institute until his death. Today, the institute is one of the most important centers in the world for the study of infectious diseases and other subjects related to microorganisms, including genetics.

After suffering several strokes that left him unable to move his left arm or leg normally, Pasteur died in Saint-Cloud, France, on September 28, 1895. He was a national hero and had received many awards and honors. His remains now lie in a special vault at the Pasteur Institute in Paris.

Linus Pauling (1901–1994)

Linus Carl Pauling was one of the most brilliant, respected, and beloved scientists of the last century. Some people have suggested that the wise character of Yoda in *Star Wars* was based on Pauling. During his life, Pauling, a chemist and biologist, received not one, but two Nobel prizes. He is the only individual ever to receive two Nobel prizes not shared with anyone else. He received the 1954 Nobel Prize in Chemistry for his research on the forces at work between molecules, and the 1962 Nobel Peace Prize for his work on ending the testing of nuclear weapons.

When asked why he became a scientist, Pauling said, "I wanted to understand the world!" To him, this understanding meant learning about life at its most basic, molecular level. His work eventually helped scientists discover the double-helix shape of DNA.

Pauling was born in Portland, Oregon, the oldest child of Lucy Isabelle Darling and Herman W. Pauling. Because his family couldn't afford to pay for his college education, Pauling worked his own way through Oregon State Agricultural College in Corvallis by teaching college classes that he had only just finished taking himself. He graduated in 1922 with a degree in chemical engineering.

Pauling then began doing research at the California Institute of Technology (Caltech) in Pasadena. He received his doctoral degree from Caltech in 1925. As a graduate student, he was one of the first people to use X rays to study the structure of molecules.

From 1925 to 1927, he received a scholarship to continue his studies of molecules with other scientists in Europe. The scientists were studying the structure of atoms and molecules. What Pauling learned in Europe laid the foundation for his future achievements.

Dr. Linus Pauling examines a model of a molecule in his lab in Pasadena, California, in May 1952.

After returning to the United States in 1931, Pauling became a professor at Caltech. He also published a paper called "The Nature of the Chemical Bond," which became a classic. In it, Pauling explained how atoms link together to form molecules. Each molecule is made up of atoms held together by chemical bonds. A bond is created by an *electrostatic* (ee-*lek*-troh-**stat**-ik) *charge,* meaning that atoms with opposite electric charges—positive or negative—are attracted to each other. Atoms with the same charges keep away from each other. Pauling was the first person to understand that bonds form within a molecule. His ideas on chemical bonding formed the basis for modern theories on the structure of molecules.

In 1936, Pauling left teaching to become the director of Caltech's Gates and Crellin Laboratories. Three years later, he published *The Nature of the Chemical Bond and the Structure of Molecules and Crystals,* which contained most of his ideas on bonds in one book.

In the 1940s, Pauling began studying proteins in blood. With American chemist Robert Corey (1897–1971), Pauling worked on the structures of *amino acids,* molecules that form the basic building blocks of all proteins. Pauling and Corey studied how amino acids form long, chainlike compounds called *polypeptides* (*pahl*-ee-**pep**-tydez). These polypeptides, in turn, link together to form proteins.

Pauling and Corey believed that many proteins are held together by bonds of *hydrogen,* one of the most common gases in the atmosphere. They found that hydrogen bonds give protein molecules a spiral, or helical, shape. Their research into proteins and their hydrogen bonds helped British biochemist Francis Crick and American biochemist James Watson in their search for the structure of DNA. Watson and Crick eventually determined that the structure of the DNA molecule is a double helix, a shape like a spiraling ladder.

Throughout the 1940s, Pauling also worked on understanding the disease *sickle-cell anemia.* Sickle-cell anemia is a painful disease caused by misshapen red blood cells. Normal red blood cells are soft, round, and flattened. Their shape allows them to move easily through the body's blood vessels. Red blood cells in people with sickle-cell anemia, however, are hard, sticky, and shaped like bananas—long and curved. Sickled red blood cells don't pass easily through blood vessels because they stick together, blocking the flow of blood and causing severe pain.

Sickle-cell anemia, which affects mostly blacks, is inherited. Pauling was the first to realize that sickle-cell disease was inherited. He also discovered that the sickle shape is caused by a flaw, or *defect,* in hemoglobin molecules. *Hemoglobin* (**heem**-uh-*gloh*-bin) is a protein in red blood cells that carries oxygen. A slight defect in the hemoglobin molecule causes the hemoglobin to form into long, stiff rods. The rods give the cells their sickle shape. Pauling's discovery paved the way for more effective treatments for this disease.

During the 1950s, Pauling became concerned about the spread of nuclear weapons. In 1958, he presented a letter to the United Nations (UN), urging an end to the testing of these weapons. *Nuclear weapons* release a tremendous amount of destructive energy in a single blast. The first nuclear weapon ever used in a war was an *atomic bomb,* called "Fat Man," dropped by the United States on the Japanese city of Hiroshima on August 6, 1945, during World War II. Three days later, the United States dropped another bomb, called "Little Boy," this time on Nagasaki, Japan. The war ended less than a month later, when Japan surrendered to the United States on September 2, 1945. Historians say that the two bombs, which killed an estimated 200,000 people, played a key role in ending the war.

Pauling felt that the world shouldn't continue testing nuclear weapons. His signature on the 1958 letter was among 11,021 signatures from scientists around the world. That same year, he published a book, *No More War,* about his views on nuclear weapons. The book warned the public about the biological danger of radiation resulting from the testing of nuclear weapons. Nuclear weapons release huge amounts of radiation. Exposure to too much radiation can cause cancer. Pauling hoped that his signing of the UN letter and his book would persuade the U.S. government to stop testing nuclear weapons.

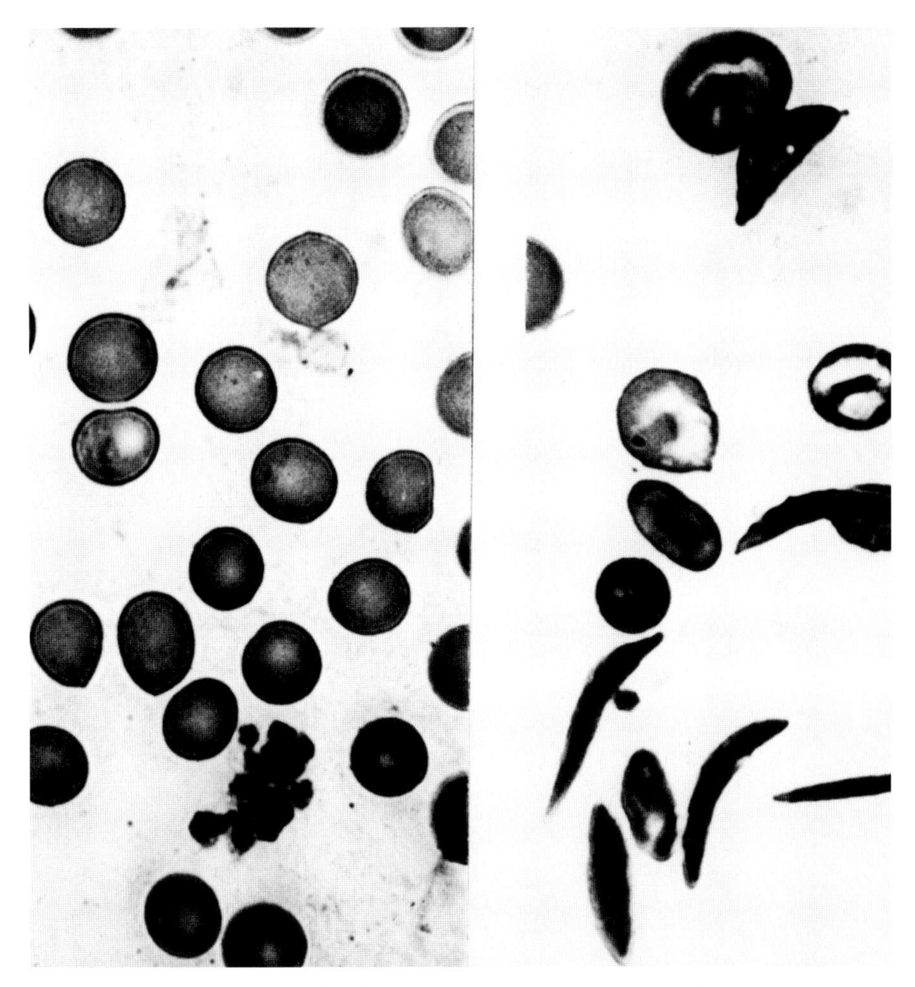

This photo shows normal red blood cells on the left and cells affected by sickle-cell anemia on the right.

In the 1960s, Pauling spent several years examining the problems of war at the Center for the Study of Democratic

Institutions in Santa Barbara, California. He used his status as a world-famous chemist to help stop nuclear testing. Pauling spoke to presidents and other world leaders about peace. Because of this work toward world peace, Pauling was awarded the Nobel Peace Prize in 1962. That year, many nations, including the United States, signed an agreement to end nuclear testing.

During the 1970s, Pauling came to believe that large doses of vitamin C could prevent the common cold and fight off cancer. In 1973, he became the director of the Linus Pauling Institute of Science and Medicine. (The Institute was originally located in California but is now located in Corvallis, Oregon.) Pauling's institute is devoted to his ideas about using nutrition and vitamins to cure disease. Pauling remained director of the institute until his death in 1994, at the age of ninety-three.

Francesco Redi (1626–1698)

Francesco Redi (fran-**chess**-ko **ray**-dee) was the first scientist to challenge the theory of spontaneous generation. As late as the fifteenth century, people still believed that maggots came from decaying meat, worms came from decaying horsehair, and frogs came from pond slime. Redi proved the theory wrong.

Redi was the oldest of nine sons born to Gregorio and Cecilia de Ghinci (deh **gin**-shee). He was born in Arezzo, Italy, on February 18, 1626. Redi's career followed in the footsteps of his father, who became the personal physician of the Grand Duke of Tuscany. Redi studied medicine in Florence, Italy, and received a degree in medicine from a university in Pisa, Italy, in 1647.

In the following years, as he practiced medicine, Redi became aware that if meat was covered, maggots didn't appear on it. Maggots, which look like little worms, are flies in an early stage of their growth. Redi pointed out a flaw in the theory of spontaneous generation. He saw that maggots grew in uncovered meat but not covered meat. His research offered the first evidence that the theory of spontaneous generation was not correct.

Italian physician Francesco Redi disproved the theory of spontaneous generation 200 years before Louis Pasteur did.

Redi's logical, scientific work wasn't enough to persuade people, however. The belief in spontaneous generation continued for another 200 years, until Louis Pasteur (1822–1895) again disproved the theory. Even at the late date of 1863, scientists still argued with Pasteur, challenging his idea that life could be produced only from other life.

Although Pasteur was given credit for finally disproving spontaneous generation, his experiments were similar to the work Redi had done centuries before. Redi died on March 1, 1697, in Pisa, Italy.

Maxine Singer (1931–)

Maxine Frank Singer, a chemist and DNA researcher, has received many awards for her contributions to science in biochemistry and molecular biology. She has also spoken out frequently about the responsibilities that scientists have toward society and about the need for math and science education in the United States.

Singer was born in New York City on February 15, 1931. She went to New York City public schools and in 1952 received a degree from Swarthmore College in Swarthmore, Pennsylvania. She received a doctoral degree in biochemistry from Yale University in New Haven, Connecticut, in 1957. Following graduation, she worked at the Institute of Arthritis and Metabolic Diseases, a division of the NIH in Bethesda, Maryland. She remained there until 1975.

At the NIH, Singer became interested in DNA and a similar molecule called RNA. She was particularly interested in the structure of RNA, studying the chemicals that break it down in bacteria.

From 1971 to 1972, Singer conducted research at the Weizmann Institute of Science in Rehovot, Israel. There, she did experiments on *simian virus 40,* a virus found in monkeys. She eventually discovered a so-called jumping gene in human DNA. Jumping genes, or transposons, are genes that move on a chromosome. The discovery of transposons on chromosomes helped scientists better understand how bacteria can withstand the effects of infection-fighting drugs. Transposons may also be involved in turning a normal cell into a cell that can cause cancer.

Stanford University professor Irving Weissman (center) chairs a meeting of a National Academy of Sciences panel on human cloning in Washington, D.C., in January 2002. He is flanked by Maxine Singer (left) of the Carnegie Institution and Mark Siegler of the University of Chicago.

Singer discovered a transposon that can cause some forms of *hemophilia* (*hee*-moh-**fil**-ee-uh), a blood disease in which the blood doesn't form clots. People with hemophilia can bleed heavily from even a small cut. Singer's research with the hemophilia transposon has led to greater understanding of this inherited disorder.

Singer was also part of a committee that organized the famous 1975 Asilomar Conference, a gathering of molecular biologists from around the world. It was held at the Asilomar Conference Grounds in Pacific Grove, California. The conference provided a way for scientists to discuss the field of genetics and its future. The conference declared a stoppage, or *moratorium,* on certain DNA studies. The moratorium continues to be followed to this day.

After the Asilomar Conference, Singer said:

> *[The conference] was important because it allowed us to take a scientific approach to the issues. One of the first things that was done was to declare a moratorium on several kinds of recombinant DNA experiments. A lot of scientists thought that was foolish, but a significant number thought there might be a serious problem with those experiments. As far as anyone knows, the moratorium was adhered to without exception all over the world.*

In 1988, Singer became the president of the Carnegie Institution of Washington, D.C. At Carnegie, Singer started programs to help the public better understand science. These programs include training elementary school teachers in science and a Saturday program for children in inner-city Washington. Singer continues to write and speak about the issues with which she is most concerned.

Robert Swanson (1947–1999)

Robert Swanson has been called "the man who invented the biotech business." He was a businessman from Menlo Park, California. Menlo Park is located in the famed Silicon Valley, known for its high-tech companies. Swanson combined his business knowledge and creativity with a willingness to take risks. The result was Genentech, the first and perhaps most successful biotechnology company in the world.

Robert Swanson, founder and CEO of the biotechnology company Genentech, poses in the company labs.

Swanson was born on November 29, 1947, in Brooklyn, New York. An only child, Swanson grew up in Florida. He graduated from the Massachusetts Institute of Technology with a basic degree in chemistry and an advanced one in business administration. He was the first graduate of that school to complete an advanced degree—a master's degree—in the fourth year of college.

After graduation, Swanson worked for a huge finance company called Citibank. In 1975, he left Citibank to join Kleiner & Perkins, a firm that helped companies just starting out to get on their feet. While working at Kleiner & Perkins, Swanson became aware of a scientific advance that he thought might make a good business.

Herbert Boyer, a researcher at the University of California at San Francisco, and Stanley N. Cohen, a researcher at Stanford University in Stanford, California, had been experimenting with recombinant DNA. The result was a new kind of organism—in Boyer and Cohen's research, a new bacterium.

Swanson wondered whether new organisms formed by recombinant DNA could be used to make medicines. He decided that if it were possible, he would give up his job to start a new company based on the science of genetic engineering.

Swanson spoke to a number of scientists working with recombinant DNA. Each one told him that the ability to make medicines from recombinant organisms was many years away. Then in 1975, Swanson asked Boyer for a ten-minute meeting. The meeting lasted much longer. The men became friends, talking for hours about using recombinant DNA technology to make and sell medicines.

By 1976, Swanson had persuaded Boyer to join him in starting a new company. Swanson also persuaded his boss, Thomas Perkins, to loan him money to start the business. The first goal of the new company, Genentech (short for *gen*etic *en*gineering *tech*nology), was to make artificial human insulin. Many people with diabetes need insulin injections to survive.

Before Swanson and Boyer came along, insulin was produced primarily from pigs. The pig form of insulin caused allergic reactions in some people, however. Genentech set about to develop a human form of insulin that would have less risk of allergic reaction and fewer side effects. In 1978, Boyer produced an artificial version of the gene that makes human insulin. The gene was inserted into bacteria, causing the bacteria to produce human insulin.

In 1985, Genentech developed artificial human growth hormone, necessary for proper growth and development of bones and other tissues. The company has also used recombinant DNA technology to manufacture interferon (an antiviral drug) and tPA (a drug used to stop heart attacks from happening).

Swanson was the head of Genentech from 1976 until 1996, when he retired. Had Swanson not realized that recombinant DNA technology could be used to make medicines, many of those medicines used every day in hospitals around the world might never have been created.

Swanson died on December 6, 1999, after a fifteen-month battle with brain cancer. He was fifty-two.

James Watson (1928–)

James Dewey Watson was one of the scientists who discovered the double-helix shape of the DNA molecule. Watson was born on April 6, 1928, in Chicago, Illinois. Because he was such an intelligent student, he entered the University of Chicago at the age of fifteen. He graduated in 1947 at the age of nineteen, when many of his friends were just entering college.

After receiving his Ph.D. in genetics from Indiana University in 1950, Watson joined the Cavendish Laboratory at Cambridge University in England. There, he met Francis Crick. Crick, along with colleagues Maurice Wilkins and Rosalind Franklin, was trying to identify the structure of DNA. Watson formed an instant friendship with Crick.

The king of Sweden applauds as James Watson receives the Nobel Prize for Physiology or Medicine in Stockholm in December 1962. Watson and Francis Crick won the Nobel prize for their work discovering the structure of the DNA molecule.

Within three years, Watson and Crick discovered that the DNA molecule consisted of two strands of atoms with the shape of a spiral. They called this shape a double helix. Franklin's X-ray photos of the molecule proved highly valuable in Watson and Crick's discovery. The Watson-Crick model of DNA caused great interest in the fields of molecular biology, genetics, and biochemistry.

In 1955, Watson joined the biology faculty at Harvard University in Massachusetts. In 1968, he became director of the Cold Spring Harbor Laboratory on Long Island in New York. That same year, he published *The Double Helix,* his account of the discovery of DNA's structure. The book became a best-seller and made Watson and Crick well-known to both the general public and other scientists almost instantly.

In 1962, Watson shared the Nobel prize with Crick and Wilkins for their work with DNA. Watson continued to head the Cold Spring Harbor Laboratory and was noted for his management abilities. In fact, he helped the lab become one of the world's leading research centers for molecular biology.

In 1989, Watson became director of the National Center for Human Genome Research, now called the National Human Genome Research Institute (NHGRI). The NHGRI is a division of the NIH. While at the NHGRI, Watson helped start the Human Genome Project.

The Human Genome Project's goal is to locate and identify all the genes in the human genome. With more than 3 billion bases to sequence, some scientists weren't sure that the project would succeed. However, Watson rallied much support for the project, which succeeded in part of its goal in 2000. In 1994, after six years as the project's director, Watson left the Human Genome Project to assume the presidency of the Cold Spring Harbor Laboratory.

In looking back over his accomplishments, Watson believes that the most important discoveries come from breaking away from what everyone else is doing and thinking. In an interview at the University of California at San Francisco in 1992, he said,

> *You know, if you're going to make the next step in a major scientific thing, no one knows how to do it so you have to, in a sense, reject your professors and say, "They're not getting anywhere, I'm going to try something else." Crick and I did that at one stage and we're famous practically because we thought that what other people were doing [wouldn't] get anywhere.... That's part of your education, to know what things won't work, and then try to get something to work.*

He added, "We were, of course, pretty lucky."

Geneticists James Watson (left) and Kary Mullis shake hands at a meeting on PCR in September 1994.

Maurice Wilkins (1916–)

Maurice Hugh Frederick Wilkins's work in discovering the double-helix shape of DNA earned him the 1962 Nobel Prize in Physiology or Medicine, along with James Watson and Francis Crick. Wilkins was born of Irish parents in Pongaroa, New Zealand. His father was a doctor.

At age six, Wilkins was brought to England to attend King Edward's School in Birmingham. He studied physics at St. John's College in Cambridge and graduated in 1938. He earned a doctoral degree in 1940 at the University of Birmingham.

During World War II, Wilkins was asked to work on the famous Manhattan Project, the attempt by a group of leading scientists in Los Alamos, New Mexico, to develop an atomic bomb. The project eventually developed the technology for the first nuclear weapons. Historians say that the two atomic bombs that the United States dropped on Japan to end the war killed an estimated 200,000 people.

Seeing the destruction that his work had helped to cause, Wilkins became a strong opponent of nuclear weapons. His opposition was not popular in the years following the war. He remains an opponent of nuclear weapons to this day.

After the war, in 1945, Wilkins went to Scotland and taught physics at the University of St. Andrew's in Fife. There, he became interested in the study of biology. He moved to King's College in London, England, in 1946, to become part of a newly formed group called the Biophysics Research Unit. Wilkins began examining DNA using microscopes and X-ray machines. He and a colleague, Rosalind Franklin, used a special X-ray technique to examine crystals of DNA. That technique helped Wilkins and Franklin find the spiral shape of DNA, and that discovery gave Watson and Crick the information that they needed to fully define the double-helix structure of the DNA molecule. Wilkins spent the rest of his career teaching and speaking out against nuclear weapons.

Dr. Maurice Wilkins studies a model of a DNA molecule following the announcement that he was to be awarded the 1962 Nobel Prize in Physiology and Medicine. Wilkins shared the award with Francis Crick and James Watson for their discovery of the structure of DNA.

Ian Wilmut (1944–)

Ian Wilmut is a familiar name to anyone who read about the cloning of Dolly the sheep in 1997. *Cloning* is the use of genetic engineering to create a group of cells, a plant, or an animal that have exactly the same genes as the original. Wilmut headed the team that cloned Dolly, the first mammal cloned from an adult cell of another mammal. Many scientists thought that this feat was impossible until Wilmut and his team succeeded.

Wilmut was born in Hampton Lucy, England, in 1944. He earned a college degree in agricultural science at the University of Nottingham in England. In 1971, he received a doctoral degree in animal genetic engineering from the University of Cambridge in England.

Shortly after earning that degree, Wilmut produced Frosty, the first calf born from a frozen embryo. Wilmut next joined the staff of the Animal Breeding Research Station in Roslin, Scotland—a facility that later became known as the Roslin Institute. There, Wilmut began the work that would eventually bring Dolly into the world.

Before Dolly, the only successful cloning experiments had used cells from a *blastocyst* (**blast**-oh-*sist*), an extremely early stage of development. Within about four days after an egg from a female unites with sperm from a male, the cells form a layered clump of cells—the blastocyst. The outer layer of cells will eventually form tissues to support the unborn child during its development. Those tissues include the *placenta,* a saclike organ that supplies the growing fetus with blood from the mother.

The cells inside the blastocyst, also known as *embryonic* (*em*-bree-**ahn**-ik) *stem cells,* eventually develop into cells that perform particular functions in the body. For instance, some stem cells become blood stem cells, which eventually become white blood cells (which fight infection), red blood cells (which carry oxygen), and platelets (which help the blood clot).

Before Wilmut, scientists thought that after embryonic stem cells had started to become other kinds of cells, they could no longer be used to clone an entire animal. Instead, they believed, for instance, that a blood cell could be cloned only into other blood cells.

Wilmut and his team of scientists had a different theory. They thought that if they didn't provide nutrients to a cell, it might hibernate. A *hibernating* cell goes into a long "sleep" in which its life functions operate only slightly. Wilmut could then transfer the sleeping cell's genes into another cell and then cause that cell to begin dividing.

Wilmut and his team set out to prove their theory. They removed the nucleus, which contains the DNA, from a female sheep's egg cell. The scientists then replaced the nucleus with a hibernating nucleus from a cell obtained from another adult female sheep. An electrical charge awakened the hibernating cell nucleus. The scientists placed the modified egg into the *womb* (woom), or *uterus* (**yoo**-ter-us), of a third female sheep. The womb is the organ in most female mammals that holds the young before birth. Wilmut's team performed this procedure 275 times before an egg finally developed normally in a sheep's uterus. That egg became Dolly, born in 1996. Wilmut's team announced her birth early in 1997.

The news about Dolly created a sensation. People around the world discussed cloning and whether scientists should continue to research cloning. Since Dolly's birth, other scientists have used similar methods to clone other mammals, including cows, mice, and cats. The ability to create clones is important because it lets scientists create identical animals that contain desirable traits, such as the ability to provide high-quality milk or resist certain diseases.

Wilmut continues to work in genetics at the Roslin Institute. He has spoken out against attempts to clone humans. He says it is much too early in the development of cloning science to even think about cloning humans. Instead, Wilmut is looking at ways of producing animals that act as manufacturing plants for proteins that can be made into medicines.

Wilmut has spoken many times about the need to continue cloning research. He says that people should continue to discuss the topic and to make sure cloning isn't used for the wrong reasons:

Dolly the sheep, the first mammal cloned from the adult cell of another mammal, stands in her pen.

Dr. Ian Wilmut, the head of the Scottish research team responsible for the cloning of Dolly the sheep, explains the process and problems since the experiment during a news conference in November 1997.

We want to stimulate...public discussion of the way [cloning] techniques might be misused.... But...we don't [want to] throw the baby out with the bathwater. There are real...benefits [to cloning], and it's important that the concern to prevent misuse doesn't also prevent the really useful benefits that can be gained from this research.

Timeline of Biotechnology

This section presents a timeline of key events in biotechnology.

DATE	EVENT
6000 B.C.E.	Ancient nomadic tribes make beer using grain that has fermented. When grain ferments, yeast in the grain breaks down sugar and produces carbon dioxide gas and alcohol. Today, beer is one of the most widely consumed beverages on Earth.
1500 B.C.E.	Ancient peoples realize that purple foxglove, a flowering plant, can heal sick people. The plant is now known to contain a substance, *digitalis,* that slows and strengthens the heartbeat. Digitalis is perhaps the most famous example of a drug that comes from plants and has saved countless lives over the years.
1000 B.C.E.	Hindus observe that certain diseases might "run in the family" for some people. They believed that children inherited all of their parents' characteristics.
Early 4th century B.C.E.	The Greek philosopher Anaximander (c.611–c.547 B.C.E.) proposes that living creatures emerged from slime. His theory became known as *spontaneous generation.* It remained the common belief about the origins of life until the great Louis Pasteur (1822–1895) proved the theory wrong.
Early 6th century B.C.E.	The Greek physician Hippocrates (c.460–c.377 B.C.E.) describes a bitter powder made from the bark of a willow tree. The powder eases pain and reduces fever. Scientists later found that the powder contained a chemical called *salicin.* This chemical was later made into another chemical called *salicylic acid.* Salicin is responsible for the painkilling properties of aspirin. Salicylic acid remains the key ingredient in today's aspirin, which is used to reduce pain, fever, and inflammation (swelling). It is also used to prevent—or even stop—heart attacks, if taken as soon as the pain of an attack begins.

1300s An Asian army catapults plague-infested corpses into an Italian trading post. Plague is a disease caused by bacteria that are normally spread by fleas. Traders in the post became infected with the bacteria and developed the disease. When the traders who survived returned to Italy, the plague spread quickly. This event is probably the earliest recorded instance of biological warfare.

Mid-1600s Anton van Leeuwenhoek (1632–1723) builds his own microscopes and makes thousands of observations about microscopic life. His observations proved critical for scientists everywhere and are still considered accurate today.

Late 1700s Robert Bakewell (1725–1795) experiments with selective animal breeding. Bakewell identified male and female sheep with the most desirable traits and allowed those sheep to mate. When livestock is bred this way, each generation of sheep possesses more desirable traits and fewer undesirable ones. Bakewell's sheep were known for their high-quality wool and tasty meat.

1796 Edward Jenner (1749–1823) develops a *vaccine* for smallpox. A vaccine is a mixture of weakened or killed disease-causing organisms, such as bacteria or viruses. When a vaccine enters a person's body, it causes the body's immune system to produce special proteins called antibodies. These can protect the person against the disease for many years. Jenner believed that vaccination could someday destroy smallpox forever. Although he didn't live to see it, his dream came true in 1980, when the World Health Organization (WHO) declared that smallpox had been wiped out throughout the world. The virus itself exists now only in a few laboratories in Russia and the United States.

1856 One of the greatest scientists of all time, Louis Pasteur (1822–1895), begins several experiments to find out where

germs come from. For centuries, many people had believed in the theory of spontaneous generation. They thought that small life forms, such as lice and fleas, were born from decaying matter. Pasteur, however, thought that germs blew around freely in the air. With this idea in mind, he later invented the process of *pasteurization*. During pasteurization, a fluid is heated just enough to kill the germs in it—what are now known to be bacteria. After being heated, the fluid is instantly covered to keep other bacteria out. Pasteurization prevents the fluid from spoiling. This process is still used today on wine, milk, and other beverages.

1859 Charles Darwin (1809–1882) publishes *On the Origin of Species by Means of Natural Selection,* a landmark book in science. In the book, Darwin suggests that animals adapt their forms over time in response to their environment. Darwin's theories immediately sparked debates that continue today.

1865 A monk named Gregor Mendel (1822–1884) publishes a short scientific paper called "Experiments with Plant Hybrids." The paper describes how traits are passed from generation to generation. Mendel believed that traits are controlled by what he called *atoms of inheritance,* which never break or bend. Today, these atoms of inheritance are called genes. Although overlooked at the time, Mendel is now considered one of the most important pioneers in the history of science.

1882 Robert Koch (1843–1910) discovers the bacterium that causes tuberculosis, a lung infection that destroys lung tissue. After Koch's discovery, researchers elsewhere began to develop better treatments for it. Koch also discovered the cause of numerous other bacterial infections. In the process, he developed a set of research rules now known as *Koch's postulates.* Bacteria researchers today continue to rely on those rules in their own research.

1885 Louis Pasteur develops the first successful vaccine for rabies. Pasteur tested the vaccine on a nine-year-old boy who was bitten by a dog. The vaccine saved the boy's life and has gone on to save the lives of countless others.

1928 Sir Alexander Fleming (1881–1955) accidentally discovers a bacteria-killing mold. The mold contains a chemical that he calls *penicillin,* which would become the most important infection-fighting drug in history.

1949 Percy Julian (1899–1975) develops *cortisone* from soybeans. Cortisone is a hormone used to treat many diseases, including arthritis (a painful joint disease), acne (a common skin disease), and systemic lupus erythematosus (a disease of the body's immune system).

1953 Francis Crick (1916–) and James Watson (1928–) discover that a DNA molecule is shaped like a double helix, a sort of winding ladder with rungs of chemical bonds between the two sides. Their description changed biotechnology forever and formed the basis for the title of a famous book about DNA, *The Double Helix.* Crick, Watson, and colleague Maurice Wilkins (1916–) were awarded the Nobel Prize in Medicine or Physiology in 1962 for their work with DNA. A fourth colleague, Rosalind Franklin (1920–1958), had performed research that proved critical to Crick and Watson's theory. Franklin died, however, before she could receive an award. Nobel prizes are only given to living scientists. Crick believes that had Franklin lived, she surely would also have received a Nobel.

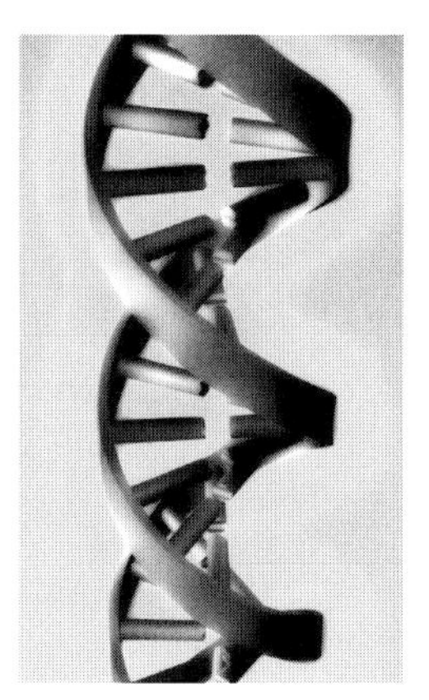

1962 Linus Pauling (1901–1994) becomes the only individual ever to receive two Nobel prizes not shared with anyone else. One of the most brilliant, respected, and beloved scientists of the last century, Pauling received the 1954 Nobel Prize in Chemistry for his work on molecules and the 1962 Nobel Peace Prize for his work on ending the testing of nuclear weapons.

1975 A conference at the Asilomar Conference Grounds in Pacific Grove, California, marks a turning point in debates about cloning. At the Asilomar Conference, molecular biologists from around the world gathered to discuss where the science of genetics was heading. They declared a stoppage, or *moratorium,* on certain recombinant DNA studies, a moratorium that continues to be observed today.

1976 Herbert Boyer (1936–) and Robert Swanson (1947–1999) create Genentech, the world's first biotechnology company. Prior to Genentech, Boyer and a colleague, Stanley N. Cohen (1935–), combined sections of DNA in bacterial cells, creating recombinant DNA. The bacteria were then able to make certain proteins. This breakthrough formed the basis of the biotechnology industry. Jeremy Rifkin, writing for the magazine *Business Week,* said the discovery was, for biotechnology, almost as important as learning to control fire was for primitive humans.

1978 Patrick Steptoe (1913–1988) and Robert Edwards (1925–) successfully perform the first *in vitro fertilization,* a technique in which an *ovum* (an unfertilized egg) from a female is united in a petri dish with sperm from a male. The fertilization led to the birth of Louise Joy Brown, known as Little Louise, the world's first "test-tube baby."

1980 The U.S. Supreme Court allows the awarding of a patent to Ananda Chakrabarty (1938–) for genetically modified bacteria. The bacteria, *Burkholderia cepacia,* can break down crude oil and is used to clean up oil spills from ships and storage tanks. Chakrabarty's patent was the first ever granted for an organism produced by recombinant DNA technology.

1994 Scientists use recombinant DNA technology to alter a tomato's genetic structure so that the tomato stays fresh longer. They call the new tomato the FlavrSavr.

1996 Ian Wilmut (1944–) successfully *clones* the first mammal, a sheep named Dolly. Cloning is the use of genetic engineering to create animals, plants, or groups of cells that have exactly the same genes.

1998 James Thomson (1958–) announces that he has isolated *stem cells* from unused blastocysts obtained from in vitro fertilization clinics. Stem cells are master cells that possess the potential to grow into any kind of cell in the body. Scientists believe stem cells hold the promise of curing a number of diseases.

2000 Scientists complete the first phase of the Human Genome Project by identifying all 30,000 currently known genes in a human cell.

Glossary

biotechnology—the use of living organisms to create or change other living things to perform new tasks or make new products

blastocyst—a cluster of early cells taken from a developing embryo at the beginning of its first cellular divisions; about the size of a pinhead; forms a key source of embryonic stem cells

chromosomes—threadlike structures found in the cells of all living things; mostly made up of rod-shaped portions of DNA

clone—an exact copy of a strand of DNA, molecule, tissue, or entire organism

cloning—the process of creating a copy of a strand of DNA, molecule, tissue, or entire organism

DNA—abbreviation for *deoxyribonucleic acid,* a double-stranded molecule shaped like a double helix that contains hereditary information for an organism

enzyme—a protein molecule that starts or speeds up chemical reactions within organisms without changing itself

hormone—one of several chemical messengers that control the organs of the body

inoculation—the process of intentionally giving people a mild case of a disease in order to protect them from a more powerful case of the disease later on

insulin—a hormone, secreted by the pancreas, that regulates the level of sugar in the blood

offspring—young organisms produced by parents

plasmid—a small, circular DNA molecule found in some bacteria and fungi that replicates on its own, independently of the organism's own chromosomes; commonly used in genetic engineering as a vehicle for transferring genes

protein—a chain of amino acids found in all living cells and necessary for life

recombinant DNA—also called rDNA; a molecule of DNA formed by splitting a DNA molecule from one organism and combining it with DNA from a different organism

vaccine—dead or weakened germs introduced to the body as a way to encourage immunity against those germs

Bibliography

Books

Gates, Phil. *The History News: Medicine.* New York: Scholastic, 1997.

Gonick, Larry, and Mark Wheelis. *The Cartoon Guide to Genetics.* New York: Harper Perennial, 1991.

Gottfried, Ted, and Julian Allen. *Alexander Fleming: Discoverer of Penicillin.* Danbury, CT: Grolier Publishing, 1997.

Harris, Laurie Lanzen, ed. *Biography Today: Scientists and Inventors, Volume 1.* Detroit: Omnigraphics, Inc., 1996.

Marantz Henig, Robin. *The Monk in the Garden: The Lost and Found Genius of Gregor Mendel, the Father of Genetics.* Boston: Houghton Mifflin Co., 2000.

Newton, David E. *Linus Pauling: Scientist and Advocate.* New York: Facts on File, 1994.

Sayre, Anne. *Rosalind Franklin and DNA.* New York: W.W. Norton and Co., 2000.

Web Sites

About Biotech *www.accessexcellence.org/AB*

Biotechnology Resources *academicinfo.net/biotech.html*

Biotech Information Committee *www.whybiotech.com*

Edward Jenner Museum *www.jennermuseum.com*

National Center for Biotechnology *www.ncbi.nlm.nih.gov*

Index

Note: Page numbers in *italics* indicate illustrations and captions.